JELMER MOMMERS is a climate journalist based in Amsterdam. He knows most people prefer not to talk or even think about climate change and that is exactly why he wrote this book. Denial and despair are not the only possible responses to the climate crisis.

Five years in the making, *How Are We Going To Explain This?* became a bestseller in the Netherlands. With this revised and updated translation, including responses to the corona-pandemic, Mommers brings his characteristic blend of realism and hope to the wider world.

'As a journalist, Jelmer Mommers has broken important stories about how we got in our current climate mess; as a thinker, he shows us there may still be some ways out, if we move with grace and speed. A fine account of where we stand, and where we could go if we wanted to!' Bill McKibben, author, environmentalist and activist

HOW ARE WE GOING TO EXPLAIN THIS?

Our future on a hot earth

JELMER MOMMERS

Translated by Laura Vroomen and Anna Asbury

P

PROFILE BOOKS

First published in Great Britain in 2020 by
Profile Books Ltd
29 Cloth Fair
London
EC1A 7JQ

www.profilebooks.com

First published in the Netherlands by De Correspondent, entitled
Hoe Gaan We Dit Uitleggen

This publication has been made possible with financial
support from the Dutch Foundation for Literature.

Nederlands
letterenfonds
dutch foundation
for literature

1 3 5 7 9 10 8 6 4 2

Printed and bound by CPI Group (UK) Ltd, Croydon, CR0 4YY

A CIP catalogue record for this book is available
from the British Library.

ISBN 978 1 78816 493 1
eISBN 978 1 78283 674 2

'Our task is to make trouble, to stir up potent response to devastating events, as well as to settle troubled waters and rebuild quiet places.'

– Donna J. Haraway, *Staying with the Trouble*

Contents

Preface 1

Part I: What's the Problem? 7

1. How We Got Into Trouble 9
2. There's Something in the Air 23
3. All Bets Are Off 37

Part II: Where Are We Headed? 53

4. It's Not Too Late 55
5. Future Scenario 1: Walls 67
6. Future Scenario 2: Forests 90

Part III: What Can We Do? 117

7. Alternatives and Opportunities 119
8. The Battle of the Century 135
9. The Power of Small Changes 154

Appendix 171
Afterword and Further Reading 178
Notes 181
Acknowledgements 215

PREFACE

I won't lie: climate change is a disaster. I've been working on it continuously for five years now and it still makes me recoil in horror. So if you hesitated to pick up this book, I completely understand.

To begin with there's the word 'climate'. The problem isn't so much its technical meaning: the average weather over a period of at least thirty years. The term was thought up to enable people to make general claims about the weather in a particular place. The Netherlands, for instance, has a more moderate, cooler climate than India. For most people and for most of our recent history the climate has been a given, about as exciting as the slow flow of glaciers or the composition of the air we breathe. Background. Fodder for experts.

But we all know that the word 'climate' currently has completely different connotations. Threat. Danger. In recent years the experts have been telling us in ever starker language that the climate is changing drastically due to human influence. Not in one place, not in the Netherlands *or* India, but everywhere at once. They're telling us that the earth is warming up, that rising sea levels are threatening coastal cities, that heatwaves are becoming more ferocious, that the global food supply is under pressure. They're telling us that continuing on our current path will almost certainly lead to worldwide catastrophes – and is already doing so.

On my computer I have a folder where I collect news and studies about the changing climate. It's an expanding invitation to despair. At least once a month another article comes along to make me think, *it's even worse than I thought!* Just as I'm getting over the shock of one extensive study stating that this century hundreds of millions of people will suffer water shortages due to melting ice in the Himalayas,[1] the next blow hits home: in the coming centuries it could get so hot that part of the cloud cover disappears,[2] resulting in further heat from the sun reaching the earth, in turn leading to an uninhabitably hot earth.[3]

No one knows if it will come to that. It's also possible that temperatures will rise slower than currently expected and that we'll adapt better than seems possible in our wildest dreams. But we have no guarantee whatsoever of those outcomes, and there's no alternative earth.

The truth of the matter is, we're in unbelievably deep shit. Mankind has never experienced the warmer climate we're heading for. Local droughts, local flooding, local extreme weather conditions – we're used to all that, but nothing in history has prepared us for worldwide climate disruption, with many consequences that are unpleasant in themselves and disastrous when combined.

We're entering completely new territory. The longer we continue on our current path, the more devastating the results.

*

But first, let's talk about the corona-crisis for a minute, because you might wonder why read a book about climate change when we are still suffering from the consequences of Covid-19. Is it really the right time to take on *another* disastrous subject? Is it necessary?

I think so. In this book, I will write about the political, technological, financial and societal forces that are giving us a shot at keeping the earth habitable. Yet right now, the strength and persistence of each and every one of these forces is tested by the outbreak of the corona-pandemic in late 2019.

The societal and economic repercussions of this crisis are only starting to emerge as I write this, yet it's clear they will be momentous. Will our economies be strong enough to continue investing in green energy, or will we fall back on old-fashioned fossil fuels? Will we return to flying and driving as if nothing were the matter? Will Covid-19 fuel nationalism and xenophobia or will it boost solidarity and cooperation? It all seems possible.

But there will be a day when this crisis, or at least the worst of it, is behind us. And as we recover, we need to focus on climate change once more. If there is one thing that the new coronavirus showed us, it's that the future is unpredictable, and that we are vulnerable. If there's one space where that insight needs to be applied, it's the climate debate.

Of course, the warming of the world and the corona-pandemic are very different crises. Though it may not feel like it right now, climate change is certainly bigger and more consequential – as this book will show you. The most important difference, however, has to do with time. Whereas the dreadful consequences of the virus outbreak could be felt in a matter of days, those of climate change accumulate much slower, over a period of years and decades, with reverberations through centuries. And while measures taken today to prevent the spread of the virus may have an effect in a week or two, those taken to counter climate change will only be felt in a decade or two. That's just one of the reasons why climate change is such a wicked problem. And why, like the corona-crisis, it so desperately needs our involvement to be addressed.

This book is designed to help you think about how we might make our societies sustainable, to show you how a revolution of sorts is already underway, and how you can help. Covid-19 regretfully reminded us of something we tend to forget: that we are connected to each other and to this planet. So we need to take care of one another and our shared home. The main argument of this book is that in the long run, those two things are essentially one and the same.

⋆

Everyone's dealing with global warming in their own way. Tom, one of my best friends, prefers to turn a blind eye.[4] Recently he said as much, when we went for coffee and talked about this book: 'I choose to look away.' He wants to get on with his life without feeling terrible about the state of the planet the whole time. He's certainly worried about the climate, but has no confidence in the government, because they're doing far too little. He doesn't trust businesses either, because in the end they always choose profit and, in doing so, often magnify the problem. And he feels he can't rely on his fellow human beings, because even people who are well aware of the facts still contribute to global warming. Then there are the deniers in politics, who proceed as if nothing were the matter, blindly marching towards the abyss.

On my bad days, I share Tom's despair. There are governments, organisations and citizens who do their utmost to limit the warming, but collectively, mankind – that's to say *we* – are still doing far too little. We're all hypocrites; believe me, I know. On the one hand I'm writing about global warming, I believe fervently that we should do more and I have made a number of adjustments in my life because I'm worried about the environment. For example, I no longer eat meat and stopped flying two years ago. But I still sometimes drive a car and while writing this book I skied for the first time in my life, on artificial snow would you believe. I'd be deluding myself if I pretended to be making a 'positive contribution' – when you look at the pollution I cause, that's clearly nonsense. At best I'm less of a burden to nature each year, recycling and composting a little more than the previous year. But that's really no reason to get all self-congratulatory. When it comes down to it, I'm still a polluter.

How do I explain this if someone in the future asks me what I did to keep the world habitable? It's an awkward question. I don't have a good answer. And I wish it wasn't there: that gnawing guilt, the fear, the despair.

So I decided to go on a quest. I went in search of ways to keep on looking without despairing. I tried to find an answer to the question of precisely how bad the climate situation is, and whether we have a credible chance of stopping global warming at all. I researched which interventions could make a difference and why.

In the last five years I've spoken with scientists and policymakers, with lobbyists and Shell employees, with activists and politicians from both mainstream and environmentalist parties. I attended climate demonstrations and industry roundtable discussions; I went to Paris for the big climate summit in 2015, and to South America to find out why Ecuador felt the need to extract oil in the Amazon rainforest. (The worldwide demand for oil, in part to fuel the plane that flew me to Ecuador, was certainly part of the answer.)

This book is an account of insights I gained along the way. I describe how we got here, how things could get worse (if we let it happen) and how they could get better (if we choose to make positive changes).

In Part I, I dive into history, because we can only move forward if we know why we got into trouble in the first place and what that trouble precisely entails.

In Part II, I describe two future scenarios. The first outlines a possible future if we continue to live our lives as we do now. The second shows what could happen if we take radical steps towards sustainability.

In Part III, I sketch how we can achieve that sustainable future. Waiting and hoping for the best isn't enough. Forces both old and new are in the process of shaping the future, and the question is which ones we make stronger.

*

In recent years I've discovered that people *can* bring about change if they act together. Not in some utopian future, but right now. Although you may not notice it much in everyday life, growing numbers of people are taking action. They don't believe that the future is something that just happens to us. And they're right: the story we'll tell our (grand)children later is one we're writing right now.

I use the words 'we' and 'our' very consciously. This book is really about all of us. It's about our collective history and our shared future. Everyone has their own position and their own ideas, but however diverse we are, in the end we all live on the same planet and share one global climate. I won't get all cheesy about it and insist that everyone hold hands, but I'll regularly write 'we' because this involves all of us.

One more thing you should know is that I'm from Amsterdam, in the Netherlands. That means some of my examples, especially in the later chapters, are drawn from there, too, where they're representative of broader developments happening elsewhere.

*

In this book I want to show that there's a way of looking at things in which despair is the start of something new, instead of a reason to look the other way.[5] There's a new story in the making, one in which the consequences of our actions add up – and every contribution is meaningful.

That's not to say that everything will be fine forever,[6] but that we have a chance to make a decisive change. The story I'm going to tell doesn't revolve around having or doing 'less' – less flying or less driving. It's about more and better – more happiness, more prosperity, better health. On my good days, I have the courage to say it out loud: everything is still possible.

This book is for all the Toms out there. For everyone who feels inclined to look away, but knows in their heart that that's not a solution. For everyone who finally wants to know precisely what's going on with the climate. For everyone who wants to make a useful contribution, however big or small. For everyone who thinks we'll never solve this but remains open to being thoroughly surprised.

Part I

WHAT'S THE PROBLEM?

Chapter One

HOW WE GOT INTO TROUBLE

Let me start at the beginning. Not with the climate, but with nature, all living things and our relationship to them.

No living being exists in isolation. You wouldn't be able to read this book without the 38 billion bacteria that live in your gut and convert food into usable energy.[1] But it's all too easy to forget those bacteria and make it through the day thinking you're an autonomous individual.

On a small scale, these feelings of autonomy are harmless enough. Gut bacteria digest my food, even when I ignore my dependence on them (99 per cent of the time). Trees and algae produce the oxygen I breathe, whether or not I feel any kind of connection to them.

But when we collectively forget how fully intertwined we are with other living beings, that's when we get into trouble.

Look at insects. Not only do they form the primary food source of many freshwater fish and the vast majority of birds, but by flying from flower to flower, insects such as bees also pollinate three quarters of all crops consumed by humans. That means they're virtually indispensable.

Agricultural chemicals are partly to blame for the disappearance of 80 per cent of insects in some areas.[2] Yet their worldwide decline seldom makes the front pages;[3] rarely is it a topic of debate in the run-up to elections. It's as if we think insects don't *really* matter, or at least not enough; as if the future of our society and the demise of these insects are somehow two entirely separate issues. Caring about 'the environment' – the natural world that's home to plants, humans and other animals – seems to be optional rather than essential.

How did this division between humans and all other living creatures come about? Why do we often see ourselves as living outside of nature, while the oxygen in our lungs, the bacteria in our stomachs and the pollinators of our food prove that we're all inextricably linked? To answer these questions, I'm going to take you back in time.

The beginning: hunter-gatherers and the natural world

Our species, *Homo sapiens*, emerges some 300,000 years back, in what is now known as Africa.[4] For much of our development, earth is going through an ice age. Thick ice caps cover the continents, like walls of ice that determine where life can and can't go.

For tens of thousands of years we live on the African savannah. Under harsh and often fluctuating climate conditions we discover how to make fire and develop our greatest evolutionary asset: as a species, we can learn from one another. We start hunting in groups and go where the food is, picking figs and hunting wild sheep.[5] During warmer and wetter periods our numbers increase; during colder and drier periods they decline again.[6]

Despite our limited number, as hunter-gatherers we immediately have a major impact on the world around us: it only takes one human with a flint or a fire stick to set a huge stretch of forest on fire.[7] Wherever people appear, large animals rapidly disappear.

That's not to say that we control our living environment. Around 60,000 years back, some of us leave Africa, some remain. But at this point we can't yet inhabit the entire earth; the polar ice caps block the passage from Siberia to North America. For the time being, the whole of the American continent remains devoid of any human trace.

*

Then there's a shift. The climate begins to change. The trigger is a number of natural fluctuations in the earth's orbit that causes the amount of solar radiation in the northern hemisphere to increase and decrease over periods of tens of thousands of years. Because of this, the earth has known ice ages and warmer intervals, the so-called interglacials, for 2.6 million years.

One such period of warming commences 21,000 years back, the first experienced by modern man.[8] The polar caps melt, causing the sea level to rise and the oceans to warm up. Over a period lasting some 9,000 years, the earth's temperature rises by at least 3 degrees Celsius.[9]

This warming comes at a good moment for man. With the polar caps retreating, we can finally conquer much of the rest of the globe. Some 6,000 years after the first signs of warming we manage to cross from Siberia to North America.[10] Another 4,000 years later – approximately 9000 BC in the Western calendar (and some 10,950 years before Elvis) – we've colonised almost the entire earth; only Antarctica remains too icy.[11]

Thanks to the warmer climate, the development of our species gathers momentum and we enter a unique period: for 10,000 years the climate will remain relatively warm and wet. More importantly, this is the most stable the climate has been in half a million years.[12] Hunter-gatherers in search of new food sources take advantage of the predictable conditions to domesticate wild plants. This is something we tried prior to the earth's warming,[13] but only manage now thanks to the dependability of the seasons.

Independently of each other, different groups of people in at least seven different regions begin to cultivate crops. Generation after generation, we toil away, with the harvest becoming a little bit more reliable each year. We start eating wheat, barley, lentils, maize and potatoes. More and more of us quit our nomadic hunter-gatherer lifestyle and take root in a settlement.

The Agricultural Revolution: the first divide

The development of agriculture changes everything for *Homo sapiens*. For the first time in our history we no longer have to run after our food. Our diet becomes less varied and less nutritious than when we were hunter-gatherers. But at least we no longer have to lug our babies around. We start having more children, in quicker succession, and with the extra mouths that need feeding comes the necessity to maximise yields on our land.[14]

This means that we stand to gain from controlling our natural environment and subjecting it to our needs. The fact that we modify our environment isn't that strange – all animals do. Beavers build dams, birds make nests, worms dig burrows.

What sets us apart is the scale on which we do it; we keep coming up with new ways of exploiting large tracts of land. We transform entire forests into fertile fields, and divert rivers so their water flows where we want it.

As a consequence, we begin to see ourselves less and less as 'one of the animals'. On the contrary: some 10,000 years back, we start *keeping* animals – sheep and goats to begin with – and promptly consider them to be our property.

Later we build fences around our settlements to protect ourselves and our harvests from predators and other unwanted intruders. Yard here, wilderness out there.

In this way the Agricultural Revolution marks a decisive shift in our relationship with nature: it leads, at least in our thinking, to a divide between our species and the rest.

Yet we don't outgrow nature. As farmers, we live in close harmony with our land and our livestock. We know that at any moment, nature can make itself felt in the shape of a crop failure, a sick pig or a devastating storm. The earth continues to determine our rhythm: there's a time to sow and a time to harvest.

But over the course of history the *idea* that we, as people, are separate from the rest of nature gains in strength. The more we impose our will on nature, the more attractive the thought that our unique qualities raise us above all other living creatures, above the natural world.

*

After the emergence of agriculture and the permanent settlements, other innovations follow in rapid succession: the wheel, paper and money among them. We see the arrival of religions, cities and major civilisations – in 3000 BC, Mesopotamia is the first urbanised society, and within a few centuries several more appear.

As these burgeoning civilisations flourish, we start seeing the first elites who live off the food that others grow for them. Kings, clergy, artists and thinkers are free to ponder the future and our place on earth.[15]

Through the ages, differing views of nature follow one another. In most religions respect for creation is a guiding principle, as is the idea of stewardship: that it's our duty to properly manage nature.

But at times it becomes a question of control rather than management, as we see in the Old Testament, which says that man must subdue the earth, so he'll have dominion 'over the fish of the sea, and over the fowl of the air, and over every living thing that moveth upon the earth'. This idea of man as ruler becomes increasingly dominant, especially in Europe.

The Enlightenment: confirmation of the divide

When the brilliant French thinker René Descartes appears on the scene in the seventeenth century, change is in the air in Europe. It all kicks off in 1543, when Nicolaus Copernicus surprises everybody with his claim that the earth orbits the sun. Suddenly, long-held beliefs are thrown into doubt. In the decades after, old ideas about nature and reality are gradually dismantled. This period in which Western scholars are trying to understand and explain the entire world will later enter the history books as 'the Enlightenment' or 'the Age of Reason'.

The young Descartes plays a decisive role in all this. A genuine all-rounder who specialises in mathematics, physics and philosophy, he comes up with remarkable theories in each of these disciplines. For example, he's the first to think of a mathematical formula to describe the way in which light is refracted when it passes through water.

But his biggest legacy is his attempt to explain 'the whole of nature'. In his quest for the foundation of all knowledge – his life's mission – Descartes radically divided man from the rest of the universe by claiming in 1637 that only humans have a 'mind'. In his view, the rest of nature is merely matter. Animals are objects, things that can't think for themselves.[16]

The line that Descartes draws between ourselves and the rest permeates Western thought to this day. It goes without saying that that's not because everybody in the seventeenth century read Descartes' philosophy and suddenly thought: yes, I'm an autonomous being who's above nature! Descartes' radical split is important first and foremost because it's such a clear *reflection* of the dominant worldview that emerged in Europe in the preceding centuries.

Descartes' contribution lies in making it seem rational to view man as above nature. After the Enlightenment, this is no longer just an idea; it acquires the status of scientific fact. If you disagree, you're unreasonable.

An idea that an erstwhile hunter-gatherer would have found utterly foolish – 'above nature, me?' – has become totally normal by the seventeenth century, at least in the urban centres of Europe. The division between 'society' and 'nature' is now well and truly established.[17]

So hardly anybody bats an eyelid when Descartes writes that man must become the 'master and possessor of nature'. He puts into words what many at the time are thinking, and doing. Western merchants and mariners are busy trying to rule the world. From the late fifteenth century onwards, European conquerors travel the globe in the firm belief that by taming 'the wilderness' they are in fact spreading civilisation. In the overseas colonies the hierarchy between man and nature is used as a template for the relationship between the colonisers and the enslaved. In the eyes of the merchants and kings, the world is a big wholesale market where they can shop at rock-bottom prices. And for centuries that's exactly what happens.

On the home front, meanwhile, the contours of modern industry are beginning to take shape – a new chapter in the ever-growing divide between man and nature.

The Industrial Revolution: man as master and possessor

Like any animal on earth, man uses energy. For thousands of years, ever since we started running after our prey on the savannah, animals and plants are our main sources of energy. Later we discover the power of wind (for sailing boats and turning mills), whale oil (used in oil lamps) and peat and coal (excellent sources of heat).

But it's not until the early eighteenth century that someone conceives of a machine that runs on coal. The idea is simple enough: bringing water to the boil over burning coal generates steam that can set a pump or turbine in motion. That motion can then move other things. Connect a steam engine to a loom, for example, and the work of dozens of human hands can be replaced by burning one wagon of coal. Eureka!

When in 1784 inventor James Watt applies for a patent on a steam engine that's light enough to use anywhere, our history takes another turn: the Industrial Revolution.

The burning of coal gives us unprecedented control over our energy supply. It's telling that the English language has a word that covers both 'energy' and 'might'. As the investor behind Watts' steam engine put it: 'I sell here, Sir, what all the world desires to have … POWER.'[18]

*

Just as we made nature serve our needs during the Agricultural Revolution, so we do during the Industrial Revolution – but on *steroids*. Now we're no longer constrained by limited horsepower and manpower. As we become less dependent on flowing rivers and on the wind, nature plays a less decisive role in the pace of our development.

The steam locomotive enters the picture, as does the steamboat. In addition to coal, we're beginning to extract more petroleum and natural gas. These three are all so-called 'fossil' fuels that were formed over many millennia. Yet we burn this ancient energy in the blink of an eye, in colossal quantities.

It's the dawn of a golden age for our species. Prior to the age of fossil fuels people's lives were often not much better than those of their parents or grandparents before them. Yet now the pace of our development soars.

But even as our grip on the natural world increases, the first cracks are beginning to appear in the underlying worldview.

Another worldview: nature as a web of life

For centuries the notion that people should subjugate nature has been at odds with another idea: that we're part of the natural world, and that we ought to treat it with care. In the West nobody has expressed this idea better than Alexander von Humboldt. Born in Prussia in 1769, he travelled the globe and became the most famous scientist of his day.[19]

Von Humboldt sees nature as a 'web of life' – and man as part of that web. 'In the great chain of cause and effect no piece of matter, no activity should be observed in isolation,' he wrote in 1807.

Unlike Descartes, he doesn't view nature as mere matter: 'Nature is a living whole,' he wrote, 'not a dead aggregate.' Everywhere we look, 'those organic powers are incessantly at work'.

Even the air is full of invisible origins of life: pollen, insect eggs and seeds. And it's the way all of this life hangs together that is of particular interest to him. 'Everything,' he writes, 'is interconnected.'

This may sound a bit hackneyed nowadays (especially after the coronavirus has just rubbed this fact in our face), but it's still a crucial insight. It forms the basis of what we later come to call 'ecology': the study of the relationships, interactions and interdependencies between living creatures in so-called 'ecosystems'. Von Humboldt stresses that we can't treat the earth as an inexhaustible resource. If we do, then at some point the entire web of life will unravel.

It's these ideas that establish the Prussian's name. He's popular with contemporaries such as Charles Darwin and Johann Wolfgang von Goethe, and more towns and species have been named after him than anybody else (his brother Wilhelm von Humboldt, also a famous scientist, doesn't even come close). The centenary of his birth in 1869 is celebrated the world over – from Moscow to Mexico City, and from Melbourne to Buenos Aires.

How can it be that von Humboldt was so celebrated in his own lifetime, but that nowadays we only know his name from street signs (if at all)? And why does he draw such radically different conclusions than the Enlightenment philosophers?

*

From an early age von Humboldt spends much of his time in nature. As a young boy he's constantly trying to escape the classroom so he can go to the woods instead. He's incredibly curious, can't sit still and, by his own account, feels as if he's always being chased by '10,000 pigs'. Von Humboldt is not the kind of person to learn about the world by philosophising from behind his desk, like his Enlightenment peers before him. He wants to learn by actually experiencing nature.

He soon outgrows Prussia. In 1799, when both his parents have passed away and he receives a generous inheritance, the 29-year-old von Humboldt sets off on his first expedition. He has persuaded the Spanish king to issue him with a passport for a visit to the colonies in South America. Later he'll travel to New Spain (present-day Mexico), the United States and Siberia as well.

Everywhere he goes, he makes notes on plants, animals and people like a man possessed. He draws landscapes, mountain ranges and starry skies. At considerable risk to his own life, he climbs Chimborazo, the highest volcano in the northern Andes, and studies the effect of poisons used by local tribes. The entire cosmos is his terrain – and he can't stop talking about it. During his lifetime, he writes over 50,000 letters to scientists, authors and politicians all over the world, and receives more than twice that number himself.

Von Humboldt's letters and books reveal that he views non-human life in an entirely different light than the Enlightenment thinkers. Having said that, as a young man he had no qualms about cutting open hundreds of frogs, lizards and mice to see if their muscles twitch when an electric current is applied to them. At least in his experiments, his thirst for knowledge seemed to justify the subjugation of nature.

But he objects to the large-scale hunger for supremacy that many of his European contemporaries display. He's a fierce critic of colonialism and of what he sees as the short-sighted domination of nature. 'Man cannot influence nature,' von Humboldt wrote in 1850, 'or convert its forces to his own, unless he understands the natural laws according to their relations of measure and number.'

Because his research shows him again and again how vulnerable life is, he pleads for a cautious approach to nature. Off the coast of Venezuela, he observes, unlimited pearl fishing has all but decimated the oyster population. In the Orinoco rainforest he sees how Spanish monks illuminate their churches with oil from the eggs of freshwater turtles, causing their number to dwindle further and further each year.

Man as master and possessor: recent invention, not fact

Von Humboldt is succeeded by a steady stream of new scientists, writers, artists, clerics, politicians and activists who all stress that man is part of a web of life. Each and every one of them points to the inextricable links between humans and other living beings.

And it turns out they were monumentally right. Just look at the bacteria in our gut that provide essential services, and look at what plants and animals do for us: they filter water, clean the air and produce food.

Yet despite the fact that von Humboldt and the kindred spirits that came after him were right, the conviction that we are somehow separate from nature prevails. Since the Agricultural and Industrial Revolutions we're adamant that we can control nature, that we can bend it to our will.

*

And why is that? Because this attitude has produced tangible results. The steam engine, for example, wasn't invented because somebody was mindfully tending to the relationships between living beings in nature, but because he spent ages tinkering with a machine that burned red-hot coal – until it finally obeyed and unleashed the unbridled power of fossil fuels.

In turn this led to unprecedented innovations, from combustion engines and plastics to fertiliser and pesticides. The world population exploded, but most people's lives became increasingly comfortable – an incredible achievement. Between 1970 and 2015 the number of malnourished people fell from 28 to 11 per cent of the global population. Likewise, extreme poverty, child mortality and the number of deaths from conflict or hunger dropped rapidly.[20]

No wonder then that we forgot von Humboldt and ignored his successors: we were fine without them. In many respects the world has never been in better shape, precisely because we've imposed our will on nature everywhere. But behind this prosperity an enormous tragedy is now unfolding.

What the global economy depends on for growth takes an ever greater toll on our living environment.[21] The rate at which other animal species are becoming extinct is now hundreds of times greater than it was before the advent of man.[22] Many surviving species have seen their numbers drop: the population size of mammals, birds, fish and reptiles has fallen by an average of 60 per cent since 1970.[23]

The indispensable services provided by nature – crop pollination, water purification and soil fertilisation – have crossed the 'safe limit' on nearly 60 per cent of the world's land surface, according to a study from 2016.[24] In the words of one of its authors, 'We're playing ecological roulette.'[25]

The corona-outbreak was a terrible example of that roulette going wrong. It's what you get when you trade live, wild animals like bats, rats, squirrels and monkeys on crowded markets: disease transmission. Likewise, by invading ever more wildlife habitats, we've exposed ourselves to infectious diseases that were once bound to the animal kingdom. SARS, MERS, Zika and Ebola – it's like we're inviting these pathogens to cross over to us. So Covid-19 was not, as

Six ways in which man is leaving his mark on earth

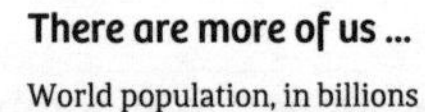

There are more of us ...

World population, in billions

... in more places ...

Domesticated land, % of total land area

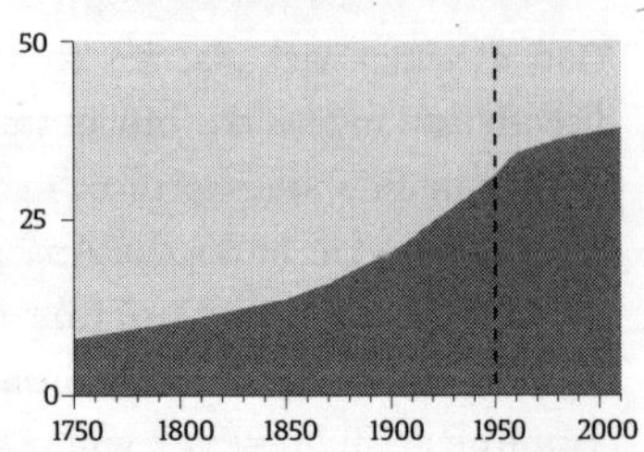

... with less rain forest ...

Tropical forest loss, % loss (area)

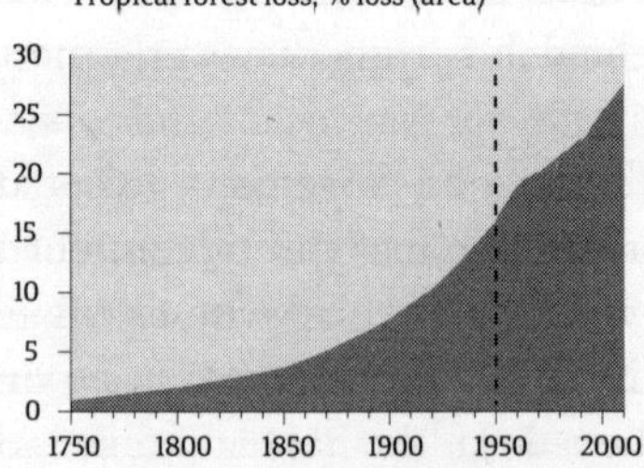

... more acidic oceans ...

Hydrogen ion, in nanomole per kilogram

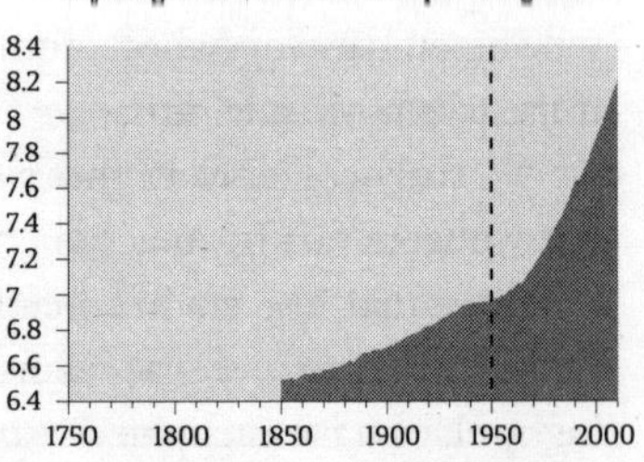

... more nitrogen in coastal zones ...

Nitrogen from (chemical) fertiliser in coastal areas, in megatonnes per year

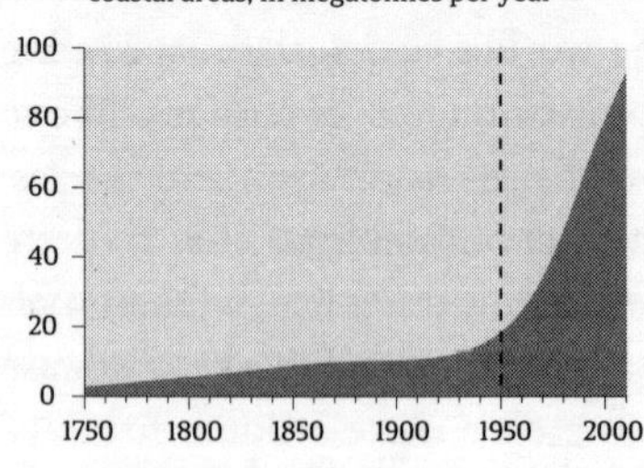

... at the expense of other species.

% decrease in mean vertebrate abundance. Index, 1970=1

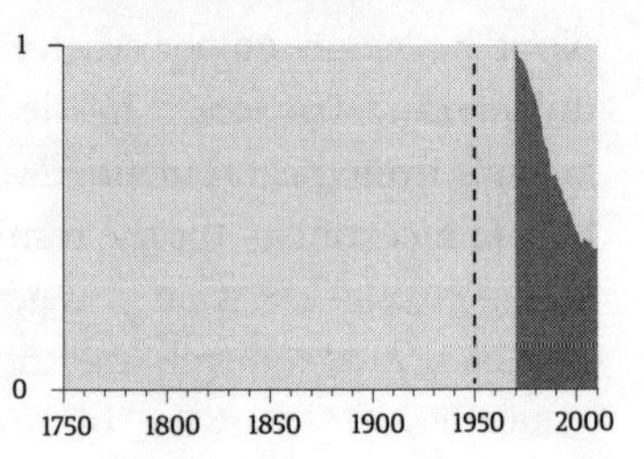

Source: WWF, *Living Planet Report 2018*

some would have it, 'nature's revenge'. It was a disaster waiting to happen – and more like it will follow if we don't change our ways.

*

Man is the deadliest species in the history of our planet. Our choices directly and indirectly impact the future of all other species.[26] We're leaving an indelible footprint – which is why we now have a geological era named after us: the Anthropocene, the Era of Man.[27]

The archaeologists of the future will find boreholes, plastic, radioactive waste and the bones of the billions of animals we consume annually. They will most likely label the 'progress' of the past few centuries as plunder: a brief phase of abundant wealth that led to general impoverishment, a *binge* resulting in a centuries-long hangover.

The archaeologists of the future will wonder why, at the start of the twenty-first century, we still hadn't unequivocally and collectively – truly collectively, that's to say, all of us together – accepted that we were facing an acute crisis. Why so many of us continued to believe that 'the environment' was nothing to do with us, while the scientists were stating in no uncertain terms that we were on the brink of a catastrophe. Why the people who *did* know failed to take sufficient action.

How are we going to explain this, when someone in the future asks us why we ignored the warnings?

A brief glance back in time gives us our first answer: because we've built our modern society on the conviction that we can be independent of nature. All the environmental crises we're now facing – from the plastic floating in the oceans to the animal species becoming extinct – are the result of that one delusion.

*

I find this insight – and this may sound strange after such a long string of depressing facts – encouraging. At least now I understand how we've ended up here, and that there's another way.

History *also* shows us that the idea of man as an independent and superior being, elevated above all other creatures, is a relatively recent *invention*, not an immutable fact. The perspective on nature that developed after the advance of agriculture and that crystallised during the Enlightenment in Europe before colonisers spread it across large parts of the globe has never been the only way to relate to all other forms of life. In fact, for the largest part of our history, it wasn't.

Knowledge about the web of life is just as much a part of our history. And nothing is stopping us from putting that knowledge centre stage again now that animals are dying in droves and the climate is changing drastically.

Chapter Two

THERE'S SOMETHING IN THE AIR

All life on earth takes place in a thin shell of air: the atmosphere. All the oxygen molecules we breathe, all the clouds the world has ever seen and all exhaust fumes from our cars float around here.

The climate within this thin layer determines the location of ice, the sea level, where we can grow food – in short, how habitable different regions of the earth are. Rapid climate change can have catastrophic effects for modern societies and for the web of life with which they're intertwined.

To explain why, in this chapter I'll take you on a tour of the main ideas and discoveries in climate science. Together they make it clear precisely what the 'climate problem' is, why scientists are so certain that the earth is warming up and why we will continue to experience the consequences for centuries to come.

Greenhouse gases and why they matter

The atmosphere begins at street level. No one can say precisely where it ends: there are gases floating around even 10,000 kilometres above the earth's surface.

That's extremely high, but 80 per cent of the gases in the atmosphere float close to the earth – up to a height of just 13 kilometres. If you could cycle for about an hour vertically into the air, you'd pass the majority of atmospheric gases. Breathing is impossible above that point: the air becomes thinner and thinner until there's nothing left and space begins.[1]

Compared with our planet's diameter – around 12,740

kilometres – the habitable part of the atmosphere is incredibly thin: a skin between the earth and the rest of the universe.

The gases that float around in this thin layer of atmosphere are crucial for life. Almost all the air – 99 per cent – consists of nitrogen and oxygen, two elements essential to all organisms. Then there's a small group of noble gases – argon, neon, helium and krypton – that contribute relatively little and simply hang around.

The rest of the atmosphere is filled with water vapour, carbon dioxide, methane and a small number of other gases, also all building blocks or by-products of life on earth. Now the gases from the latter group, which make up less than 1 per cent of all air,[2] accomplish something that makes them especially useful *and* dangerous: they trap heat in the atmosphere.

*

The Irish scientist John Tyndall discovered how the so-called 'greenhouse gases' work in 1859. He built on the discovery in 1800 of infrared radiation, invisible thermal radiation emitted by the earth's surface when the sun's rays fall on it.

As early as 1820, scientists suspected that the air around the earth somehow retains this thermal radiation.[3] Tyndall was one of the first to discover precisely how that works.[4]

In his experiments he succeeded in demonstrating that gases such as CO_2 *absorb* thermal radiation in the atmosphere, and subsequently emit it in all directions, including *back* towards the earth.

So these gases form an isolating blanket which ensures that the atmosphere can retain warmth. Thanks to this 'greenhouse effect', since the end of the last ice age the average temperature on earth has been a pleasant 15 degrees Celsius. Without greenhouse gases it would be deadly cold: an average of minus 18 degrees.[5]

So it's a good thing that there are greenhouse gases in the air: they ensure that we don't live in a world of snow and ice. At the end of the nineteenth century, Tyndall already suspected that the concentration of CO_2 gradually fluctuates over periods of thousands

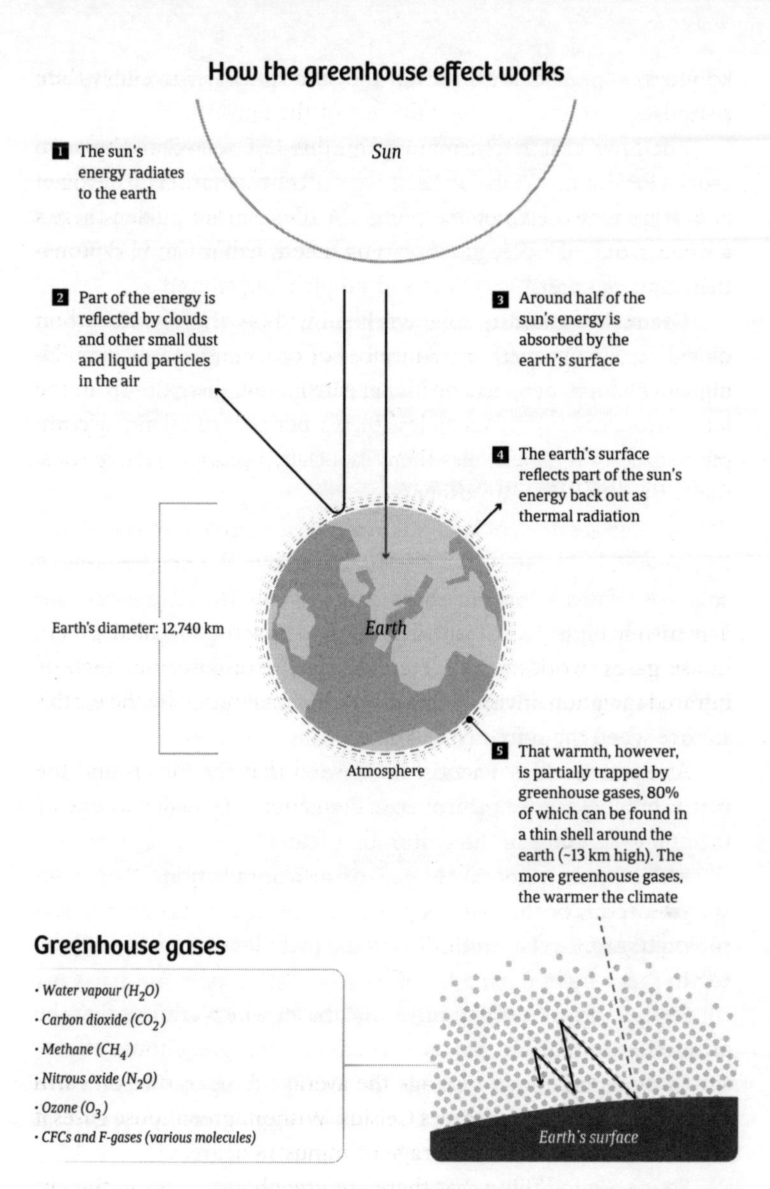

How the greenhouse effect works
Sun
1 The sun's energy radiates to the earth
2 Part of the energy is reflected by clouds and other small dust and liquid particles in the air
3 Around half of the sun's energy is absorbed by the earth's surface
4 The earth's surface reflects part of the sun's energy back out as thermal radiation
Earth's diameter: 12,740 km
Earth
Atmosphere
5 That warmth, however, is partially trapped by greenhouse gases, 80% of which can be found in a thin shell around the earth (~13 km high). The more greenhouse gases, the warmer the climate
Greenhouse gases
• Water vapour (H_2O)
• Carbon dioxide (CO_2)
• Methane (CH_4)
• Nitrous oxide (N_2O)
• Ozone (O_3)
• CFCs and F-gases (various molecules)
Earth's surface

Source: IPCC, *Fifth Assessment Report* (2013), Working group 1, Chapter 2, p. 181

of years, in parallel with the ice ages and the warmer intervening periods.

But how exactly? Increasing numbers of scientists began to wonder about this in the wake of Tyndall's observations. They kept collecting new pieces of the puzzle. Analysis of air bubbles in old ice cores, for example, has shown us how warm it was in different periods in the past.[6]

Gradually scientists have worked out how the various pieces of the puzzle fit together, and we've begun to understand why the climate changes naturally and what role greenhouse gases play.

How people prevented a new ice age

During ice ages, the cold periods in which ice caps cover a large part of the world, the intensity of solar radiation in the northern hemisphere is relatively low, and the quantity of CO_2 stored in the oceans is relatively high.

Warmer periods begin when the solar radiation in the northern hemisphere gradually increases due to shifts in the earth's orbit. The ice caps then begin to melt and the oceans warm up.

The warmer the seawater becomes, the more CO_2 it releases, just as bubbles from a glass of soda disappear faster when you place it in the sun. Once in the atmosphere, the CO_2 released reinforces the warming set in motion by the extra solar radiation.[7]

As a result of this natural process, 11,650 years ago, our warm period began. At the time, around 260 particles of CO_2 were floating in every million particles of air (the unit is 'ppm' or 'parts per million'). It was pleasantly warm, and the climate was stable enough to develop agriculture.

In our long march towards modernity we then accomplished something unique: *we have, for the time being at least, prevented the next ice age.*

That's profoundly odd! How did we do that?

If we hadn't cranked up the CO_2 concentration, the earth would now be on its way to a new ice age.[8] The planet currently receives the same amount of energy from the sun as during the height of the last ice age (i.e. relatively little).[9] And without human intervention the CO_2 concentration would have slowly dropped over the last few thousands of years.

Compare it with the way the liver breaks down alcohol in the blood. The earth eliminates high CO_2 concentrations by several processes, the most important of which is *the weathering of stone*.

This weathering process is extremely slow and powerful: CO_2 dissolved in rainwater reacts with rock, washes down rivers and out to sea, and is used by organisms to build shells or skeletons. In the end, when those animals die, the C molecule from CO_2 – the carbon – ends up on the ocean bed. Over a period of tens of thousands of years, even the highest mountains crumble, and during that 'erosion' CO_2 disappears from the air.[10]

In short, if you wait long enough, nature eliminates the high concentrations of CO_2, because year after year more CO_2 disappears from the atmosphere. This results in cooling, the growth of ice caps, and in the end – thousands of years later again – a new ice age.[11]

But at this moment we're actually *not* on our way to a new ice age. Measuring stations all over the world show that the earth is now a degree warmer than before the start of the Industrial Revolution. How is that possible? How have we managed that?

The Agricultural Revolution: Warming 1.0

The current warming began with deforestation. Carbon is stored in every tree, and in the soil of a healthy forest. But when we cut or burn down a forest, a large proportion of that carbon is released – in the form of the gas CO_2.

The opposite happens when a plant or forest grows: CO_2 from the air is stored and used by the plant or tree as a building block. The cycle of life in a nutshell. But when more forest disappears than grows over a protracted period, the concentration of CO_2 in the atmosphere rises and it gets warmer.

In 2003, American climatologist William F. Ruddiman was the first to come up with this hypothesis: that our species was already affecting the climate when we began to cut down trees on a large scale to make way for farmland.[12]

His claim chimes with the reconstructions: around 7,000 years ago the CO_2 concentration in the air certainly began to rise, from 260 ppm at the start of our warm period to 280 ppm just before the start of the Industrial Revolution.

The concentration of methane (CH_4), another greenhouse gas, also rose 5,000 years ago. According to Ruddiman this was due to the rice plantations that people were establishing around that time, which involved placing areas of ground under water, causing the plants to rot. Rotting vegetation and rice plants emit methane.[13]

It's a mean gas: it retains a good eighty times as much warmth as CO_2,[14] and is responsible for at least 17 per cent of the current warming.[15] The only plus side of this gas is that methane remains in the air on average for 'just' nine years before breaking down. But the reaction that causes the methane to disappear, sadly, also produces CO_2.

*

When Ruddiman published his hypothesis in 2003, it was still controversial. But in recent years, many scientists have put their weight behind his claim, and the evidence is mounting.[16]

Gradually consensus is gathering around one of the most fascinating facts of our history: without knowing it, 7,000 years ago we began postponing the onset of a new ice age.

Over the last few thousand years, extra warming as a result of emissions from farmers has compensated for a gradual decrease in solar radiation in the northern hemisphere. The result has been a relatively stable climate, stable enough for the development of modern civilisations, with their amphitheatres, universities and ultimately even coastal cities with millions of inhabitants.

Of course, there were always fluctuations. Droughts could raze entire civilisations to the ground – that's how Mesopotamia and the Maya civilisation met their end. But those local variations always took place within a relatively small bandwidth. For 10,000 years the average temperature on earth hardly varied – a maximum deviation of 0.7 degrees Celsius from the average, that was it.[17] These conditions turned out to be optimal for us in developing from hunter-gatherers to the people we are now.

If we'd stopped at the relatively limited CO_2 emissions of before the Industrial Revolution, the consequences would probably have been favourable. Call it warming 1.0, the benign version. The result would have been a pleasantly stable climate. Enough CO_2 and methane in the air to hold back a new ice age. Good job.

But we overshot.

Fossil fuels: Warming 2.0

When the Industrial Revolution came along we started to use fossil fuels, in gigantic quantities. This resulted in warming 2.0 – more, faster, more dangerous.

Why is the use of oil, gas and coal so risky for the climate? Let's go back one step: fossil fuels are produced when the remains of trees and plants, plankton and other organic material is pressed together in the earth's crust. This happens under extremely high pressure and over extremely long periods – thousands to millions of years.[18]

In this way the CO_2 which plants remove from the air as they grow disappears as carbon in the ground. It's safely stored. Alongside the weathering of rocks – crumbling mountains that drain away carbon – this is the second most important way in which nature gets rid of CO_2.

But this safe storage is negated when people dig up and burn these fuels – because during burning the carbon is converted back to CO_2. Between 1750 and 2019, in a short 270 years, the CO_2 concentration rose by a good 50 per cent due to our actions – from 280 ppm to 410 ppm. That's crazily fast growth.

If you compare the pace at which we're emitting greenhouse gases with previous climate changes in the earth's history, you see clearly how exceptional our impact is. During the latter part of the last ice age, for example, the CO_2 concentration over a period of 7,000 years rose from 190 ppm to 260 ppm at the start of our warm era. That's a rise of around 0.01 ppm per year. The rate of change we set in motion from the Industrial Revolution is around 0.5 ppm per year. So we've been moving at *fifty times the speed.*[19]

If slowly flowing lava is the pace of nature, we're a skydiver in freefall.[20] In fact, from 1998 onwards we have been moving even faster, raising the concentration at a rate of about 2.1 ppm per year. That's 210 times the pace of the last natural climate change. Relative to the slow-flowing lava, we're a supersonic jet breaking through the sound barrier.

*

CO_2 is the main greenhouse gas we emit, causing almost two thirds of the current warming.[21] But it's not just CO_2: the concentration of methane, nitrous oxide and other greenhouse gases has also risen significantly in the last 150 years. In the 66 million years of geological history for which we have somewhat reliable reconstructions, this pace of climate change is unprecedented.[22] Compared with 1750, we are now heading for an average global warming of 3 degrees Celsius by the end of the century.[23]

That sounds like a small change – the local temperature fluctuations between day and night are rather larger. But such variation in the average global temperature will have gigantic consequences.

Don't forget that the difference between a world of snow and ice – the last ice age – and the beginning of our comfortable, warm era was 'only' 3 to 8 degrees. The 3 extra degrees of warming we are now in danger of causing may take us to a future in which large parts of the earth are uninhabitably hot, just as large parts of the earth are uninhabitably cold in an ice age.

Compare it with your body temperature: a constant 37 degrees, with a maximum of half a degree of variation. Our bodies function within this tiny margin. Outside that, hypothermia, fever, permanent brain damage and death lie in wait.

A slight rise or temporary chill: we've experienced those in recent history. Constant high fever? That's new. And dangerous.

We tend to notice the importance of balance only when it's gone.

(Un)certainty in climate science

How certain is all this? Like any science, climate science has its uncertainties. We don't know how warm it will get in the coming centuries. It depends, among other things, on the quantity of greenhouse gases we emit in the coming decades. Specific consequences in specific places are difficult to predict.

And there is a great deal that scientists don't yet know. How cloud formation in a warmer climate might exacerbate or mitigate the warming, for example. The recent research I mentioned earlier – the prediction that a proportion of the clouds might disappear – suggests the direction things might move in, but the last word has yet to be said on the subject. And how innovatively people respond to the warming in the coming decades of course also remains unpredictable.

But since the nineteenth century, when Tyndall started his experiments, a great deal of certainty and consensus has gathered on the base principles of climate change.

An important reason for this is that researchers have consistently improved their measurements. In 1958 the American Charles David Keeling installed a CO_2 monitoring station at an observatory in Hawaii. The measurements recorded there since then give a reliable picture of the average quantity of CO_2 in the atmosphere.[24] The 'Keeling curve', named after him, shows how the CO_2 concentration is constantly creeping upwards.

Other monitoring stations, spread over the entire planet, show that the worldwide temperature moves slavishly in step with the quantity of CO_2 in the atmosphere, precisely as you'd expect.

In the 1960s, clever nerds came up with the first modern climate models. In virtual simulations of the global climate, too, the mercury rises as you raise CO_2. Oil companies were quick to confirm what was going on: the American giant Exxon concluded straightforwardly on the basis of its own research in 1977 that the world was warming due to the burning of fossil fuel.[25]

From the end of the 1970s, climate science has continued to mature, and the conclusions have become increasingly robust. Governments have started to worry more and more. In 1988 that led to

Correlation between CO_2-concentration and global temperature

CO_2 concentration since 1880

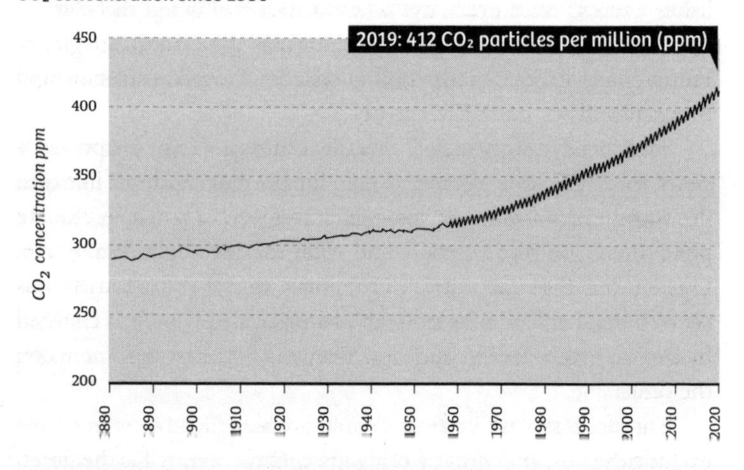

Source: CO_2 data before 1958 from analysis of ice cores, from 1958 onwards from measurements at Mauna Loa

Temperature rise compared with average in 1880-1910

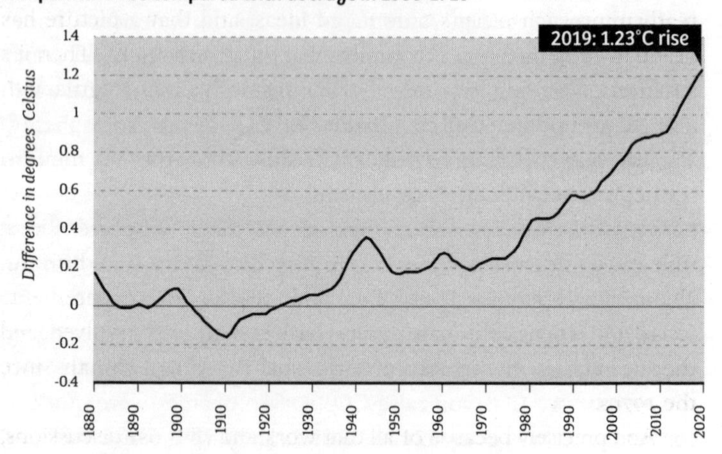

Source: NASA, *GISS Surface Temperature Analysis (GISTEMP)*

The zigzag in the CO_2 measurements from 1958 shows the effect of the seasons: the CO_2 concentration drops to a minimum in the summer because growing plants in the Northern Hemisphere absorb much of the CO_2, and rises in winter to a maximum because at that point the trees are dormant.

the foundation of the Intergovernmental Panel on Climate Change (IPCC), a scientific organisation of the United Nations which publishes a report once every five to six years, summing up *all* climate science published in peer-reviewed journals in the previous years. Hundreds of experts are invited to collaborate on each report and thousands of studies are analysed.

By critically reviewing all scientific claims made in the previous years, the IPCC experts can make up the balance. A report like this, produced every five to six years, indicates precisely which climate phenomena are most certain, and what requires further research. One of the most recent IPCC reports, published in 2018, was written by 91 researchers from 40 countries, and was then checked by 489 experts from 61 countries. They made 12,895 comments on the draft.

The process isn't perfect; people are fallible.[26] We can't even exclude the possibility that worldwide climate research is the greatest piece of groupthink in all of history: that the thousands of climate scientists publishing new studies each year are constantly reaffirming each other's unfounded ideas, and that a picture has gradually emerged that is completely at odds with reality. Then it's a matter of waiting for a new Galileo to smash all the dogma with a single groundbreaking new theory.

But that's *extremely* unlikely. Because the basis of modern science is critical analysis of each other's work, continual debate, perpetually separating sense from nonsense. Over the years all possible causes of the current warming have been held up to the light. The instruments have been tuned and retuned, all measurements tested and retested. Climate models have continuously evolved, and they've been predicting the current global warming correctly since the 1970s.[27]

And precisely because of all that work and all those discussions, because of the comparison of different evidence and different measurements, the robustness of climate science has grown enormously. We're at the point where 97 per cent of experts – the authors of peer-reviewed articles in scientific journals – agree: human emissions of greenhouse gases are the main cause of current global warming.[28]

No other factor can explain why the climate is now changing so dangerously fast. The statistical chance that rising temperatures are coincidental – and not due to human actions – is smaller than 1 in 3.5 million, or 0.00003 per cent.[29]

The tiny group of people who still doubt this are loud, but have almost never studied the climate themselves. Acting like there's genuinely still scientific 'debate' as to the causes or seriousness of the current global warming is pure ignorance. Or deception.

Why CO_2 will haunt us for centuries

If only the self-appointed sceptics were right, I sometimes think. When I was growing up in the 1990s, there was a real sense of optimism about the future in the West. When we looked ahead, we saw a road growing ever wider – that was our image of progress.[30] I had a Nintendo, an Eastpak rucksack and ninety-nine problems but 'the climate' wasn't one.

I'm probably just being nostalgic, but I'd love to recover the carefree feel of that time.[31] I'd like someone to give me a valid argument for ignoring the whole climate debate. But that seems impossible. The problem now is that our CO_2 emissions will haunt us for centuries. Because those emissions are so absurdly high.

To give you an idea of the scale: people now dump around 42 billion tonnes of CO_2 into the atmosphere annually.[32] The weight alone is almost unimaginable. One tonne is 1,000 kilograms. No object on earth weighs 42 billion tonnes. If you were to put all 31.5 million cars in the UK on scales, they would only weigh 0.06 billion tonnes. So every year we blow an inconceivably heavy object into the air in the form of trillions of molecules of CO_2.

We reach this immense volume because CO_2 is emitted wherever fossil fuels are burned: it emerges from chimneys and cars, factories and aeroplanes, power stations and container ships. Three quarters of all emissions are the consequences of these things.[33] The burning of coal causes the most CO_2 emissions, after which comes petrol, then gas. And some 5.5 of the 42 billion tonnes of CO_2 goes into the air due to the loss of forests.[34]

A *proportion* of all this CO_2 is reabsorbed each year by growing plants and trees and by the ocean. Every year these natural 'sinks' remove some 21 billion tonnes of CO_2 from the air.[35] That's around half of our emissions – an enormous amount.[36] And the good news is that in recent decades these sinks have taken up more CO_2 because there's more CO_2 in the air.[37] It's logical: lots of plants grow faster at higher concentrations of CO_2.

Sceptics often point to this to claim that CO_2 is good for life on earth, and they're partially right. Life on earth can't function without CO_2 (I feel like saying *duh* here).[38]

Parts of China, Russia, Europe, India and North America have become substantially greener in recent decades.[39] That's partially thanks to the higher concentration of CO_2,[40] and partly to reforestation, agriculture and a longer growth season.[41] But this positive effect is partly negated by cutting down forests in tropical regions,[42] by drought,[43] and by the degradation of forests in northern regions.[44]

In any case, an airbag isn't enough to prevent an accident. However hard the forests and seas work to suck up our emissions, they can't single-handedly contend with that enormous volume of 42 billion tonnes per year.

We can try to accelerate the process of breakdown of CO_2 from the atmosphere, for instance by planting new forests on a massive scale. But even these new airbags will eventually fill up again – their potential to moderate global warming is temporary, and limited.[45] The concentration of CO_2 will therefore continue to rise, and that will only come to an end when we stop adding any extra greenhouse gases to the atmosphere at all.

Even if we could wave a magic wand and end all CO_2 emissions tomorrow, in a thousand years' time 15 to 40 per cent of the CO_2 we have produced will still be in the air.[46] It will be thousands of years before nature has eliminated all those greenhouse gases, the same amount of time it takes for radioactive waste to become harmless.[47]

Technology cannot 'solve' this for us: no technology exists that can scrub billions of tons of CO_2 from the air every year. People are working on it: a handful of research teams around the world are building CO_2 filters, but the scarce successful experiments at most succeed in fixing a few tonnes of CO_2 per day. If we wanted to apply this technology on a grand scale in the future, the entire planet would have to be covered in CO_2 extractors, effectively gigantic air washers. At the current state of technology that would get rid of 1 per cent of the annual emissions at a cost of $400 billion per year.[48]

Storing that CO_2 underground would cost even more. And not only that, we'd have to go on doing it for centuries to compensate for our emissions. Not particularly attractive, and not feasible in the coming decades.[49]

A much easier solution? Stop producing greenhouse gases. Unless we do, the consequences will pile up.

Chapter Three

ALL BETS ARE OFF

How can winters be this cold when the world is getting warmer?

In January 2019, when parts of the US were hit by snowstorms so bitter and icy they wouldn't have looked out of place in the Arctic, the US president tweeted, 'What the hell is going on with Global Warming? Please come back fast, we need you!'

It's easy to sneer at Donald Trump's remark: he's confusing the weather (current meteorological conditions) with the climate (the average over a thirty-year period).

But if that's the case, then why is an extremely hot summer often cited as a symptom of global warming, yet extreme cold doesn't count as 'evidence' against?

Underlying sceptical tweets such as Trump's are serious questions about the relationship between the weather and the climate. Now that the earth's climate is warming up, the key question is what exactly is changing and what are the consequences.

Extreme heat

Since records began in 1880, the five hottest years worldwide have been the past five (2015–2019). Since 1980, every decade has been warmer than the previous one.[1]

All seasons are now warmer than they used to be. The three coldest summers in the northern hemisphere occurred in 1904, 1903 and 1913 (in that order). And the coldest winters there? They were in 1893, 1911 and 1917.[2]

It's no coincidence that the years within these sequences are so close together: warming is the trend. That's why we're totally justified in attributing a hot summer or the umpteenth heatwave to climate change. A locally cold winter, such as the one in the United States, is the exception that proves the rule – *the world is warming up*.

Most people are quick to adapt. We often don't even notice the changes in the weather. For each generation, the climate feels 'normal'. Instead of comparing this year's seasonal temperatures with a set period, we're inclined to measure them against last year's or at most against our memories of childhood summers and winters.

In fact, a recent study suggests that we accept extremely hot weather as normal within as little as two years. And so we fail to notice the trend that weather stations around the world are recording so clearly.[3] Or perhaps we do notice and simply don't care.

But other consequences of warming are harder not to worry about. All over the world heatwaves are happening more often. They last longer and are hotter too. This has numerous negative knock-on effects. Although nice warm weather is wonderful, we sleep less well when it's *too* hot, we grow more irritable faster and we're also less productive at work.[4] People labouring outside have to take more frequent breaks. Healthcare costs increase, for instance because on hot days air pollution can become trapped in the city and more people experience breathing difficulties. In the long term, more heat can also result in mosquitoes that spread malaria and dengue fever extending their habitat, thus potentially spreading their diseases far beyond the tropics: in the event of unbridled warming, a billion more people could contract malaria and dengue fever by 2080.[5]

If the heat isn't too extreme, most people will be able to handle it. Those who are most at risk, such as the elderly and the infirm, can remain inside and switch on the air-conditioning (which, incidentally, will raise outdoor temperatures even further – that's where the indoor heat is dumped).

But if the mercury rises too much, combined with an increase in humidity, all those who venture out become vulnerable to the heat. We can only expel body heat by sweating. When the humidity is too high, that's no longer possible – our sweat doesn't evaporate. At that point death isn't far off: at a temperature of 35 degrees and a relative humidity of 90 per cent even superhumans in the shade will die within hours. Visit the city of Karachi in Pakistan during a heat-wave and you'll know what I'm talking about. Like being trapped in a hot car.[6]

Luckily such a potentially lethal combination of extreme humidity and soaring temperatures is still quite rare. But unless we curb our emissions, these kinds of conditions will become increasingly common in the century to come. By 2100, three quarters of the world's population would be exposed to oppressive, muggy heat for some twenty days a year.[7] In those conditions, 'anyone could be at risk of heat-related illness or even death'.[8]

Floods and droughts

One of the most significant consequences of global warming is that water moves very differently around the planet. A warmer climate will see more water evaporate and more rain fall. The result: an upsurge in both the number of droughts and the number of extreme downpours.

The UK serves as a good illustration here – its climate is becoming (even) wetter. In the past decade, UK summers have been 13 per cent wetter compared to the average of 1961–90. Rainfall during winter is up 12 per cent,[9] meaning that extended periods of extreme winter rainfall are now seven times more likely.[10] And the amount of rain that falls on extremely wet days has increased by a staggering 17 per cent.[11]

You can blame climate change for that, because warm air can hold more water vapour than cold air – 7 per cent for each additional degree of heat. More water in the air means more precipitation when it rains. And when torrential rainfall and severe weather happen more frequently, the chance of flooding increases.

Alongside these wet conditions in one location – or in one season – we find drought in another. When the earth's surface is warmer, water evaporates faster – just like wet clothes dry quicker in the sun. In hot regions and during dry seasons, drought will intensify if the temperature continues to rise. Areas that are already seeing very little rain, such as sub-Saharan Africa, will almost certainly get even less. This could result in fresh-water shortages for millions of people – in fact in many places it already does.[12]

When forests and other natural areas dry up, the risk of wildfires escalates. That's already happening too: during the intensely hot summer of 2018, for example, dozens of forest fires were raging north of the Arctic Circle in Sweden. It wasn't 'the climate' that started these fires – lightning strikes or careless people usually do – but the more intense heat did create the conditions for fires in places where they were previously unheard of, or at the very least, less severe.[13]

In recent years dozens of people have lost their lives in extreme wildfires in California, Greece and Australia. Many others have been forced to remain indoors, with their doors and windows closed, because the air quality in the surrounding areas was so poor.[14]

These changes in the weather and these events aren't isolated cases. They can be directly linked to global warming – it has made them *more likely*.

Hot seas

Like the weather, the seas are undergoing a metamorphosis, in part because they absorb much of the CO_2 we emit and become more acidic as a result. In turn, this impacts the ability of shellfish such as prawns and oysters to grow their shells. Their numbers are likely to fall in the future.

The seawater is also warming up year-on-year – 2019 was the warmest year for the world's oceans since modern records began.[15] In recent decades more than 90 per cent of the extra heat that's trapped in the atmosphere as a result of our emissions has been absorbed by the oceans. By doing so the seas are providing a wonderful service: they mitigate the effects of our emissions. But warmer seawater has several downsides.

Firstly, tropical coral reefs are dying due to the higher seawater temperatures. These corals are the most colourful, most complex and most efficient ecosystems the earth has ever known. Although found on only 0.1 per cent of the seabed, they form the breeding ground and nutrition for a quarter (!) of all marine creatures. Hundreds of millions of people depend on the fish sustained by tropical coral reefs, but due to our emissions there's a very real risk that these reefs will disappear completely within the next one hundred years.

Warmer seawater also means that hurricanes and typhoons are now more destructive than ever, as they derive their devastating force from the sea – for a storm, warmer water means more available energy. This is also where the extra water vapour in the hot air I mentioned earlier kicks in: at higher temperatures, storms cause more rainfall, thus fuelling the risk of flooding. If we continue to emit greenhouse gases at our current level, storms will be causing much more damage in the centuries to come.

Disappearing land

The best-known consequence of global warming is the rise in sea level. When seawater warms up, it expands. This is the first explanation for the fact that sea levels globally have risen by an average of 16 centimetres since 1900.[16]

The other explanation for the sea-level rise is melting ice sheets. Due to the higher temperatures glaciers are retreating almost all over the world while the ice caps in Antarctica and Greenland are melting at a brutal rate.

The rate at which Antarctica is losing ice has more than tripled over the past fifteen years.[17] It's taken even the scientists by surprise, and unfortunately this seems to be the trend: it looks as if the ice on earth keeps melting a little bit faster than the experts had anticipated,[18] even though the globe has witnessed 'only' 1 degree of warming since 1880. The Antarctic is now suffering a net loss of more than 200 billion tonnes of ice per year – the equivalent of 200,000 Olympic swimming pools a day.[19] Because some of this ice once covered land, water levels rise when it ends up in the sea.

(A common misunderstanding is that the melting of the Arctic sea ice also adds to the rise in sea level. That's incorrect. Just as a melting ice cube in a drink doesn't cause a glass to overflow, neither does melting sea ice raise water levels.)

Unless we curb emissions, sea levels are expected to rise by 0.6 to 1.1 metres by 2100, according to the latest IPCC report on the topic.[20] But it's very uncertain how fast Antarctica will melt.[21] A rise of 2 metres this century 'cannot be ruled out', the IPCC notes.

That's excruciatingly bad news for the 65 million people who inhabit low-lying island states – at some point in the coming centuries their homes will be swallowed by the sea. Even rich countries like my own may face difficulties adapting. Although the Netherlands practically invented coastal protection, in 2016 the Dutch meteorological institute (KNMI) warned that 'we can no longer exclude the possibility that unrestrained climate change will lead to unmanageable rises in sea levels, which present an impossible task for the Dutch coastal defence system.'[22] If the Dutch can't cope, that's saying something.

Together, Antarctica and Greenland have enough ice for a 65-metre rise in sea level. But there's no need to get your inflatable ready just yet: it's expected to take at least 10,000 years for *all* of this ice to disappear,[23] if it happens at all.[24] Ten millennia – that's a timespan that borders on the unimaginable.

Somewhat easier to get our heads round is the following estimate: if today's warming trend continues, at least twenty-five countries stand to lose 10 per cent of their surface area in the next two millennia.[25] Hong Kong, Shanghai, Tokyo, Jakarta, Hanoi, Kolkata, Mumbai and New York – all of these metropolises are in the danger zone. If we fail to call a halt to our CO_2 emissions, millions of people will have to relocate in the future.

But, like nearly all of the effects of global warming I'm mentioning here, for many people sea-level rise isn't a future problem. Every year, some 24,000 people are leaving Vietnam's Mekong Delta because advancing saltwater is making agriculture impossible.[26] At spring tide, the streets of Miami are flooded, while Louisiana 'loses a football field's worth of land every hour and a half'.[27]

And President Trump, who by his own admission doesn't 'believe' in climate change, has had a wall of hay bales erected on the borders of his Irish golf course – to stop the seawater as it inches ever higher. In the planning application, Trump's golf business explicitly referred to scientific studies on global warming and rising sea levels.[28]

Crop failures

If there's one sector that's ill-prepared to deal with the consequences of global warming, it's agriculture.

Like us, our main food crops – wheat, maize and rice – are used to the relatively stable climate we've enjoyed for the past 10,000 years. But at higher temperatures – and especially during heatwaves – yields diminish. That's why European farmers struggled so badly during the summer of 2018, the hottest on the continent in over a century.

Likewise, the 'granaries' of the world can no longer pretend it's not happening: each degree of local warming reduces China's wheat harvest by 3 to 10 per cent.[29] The reason is simple: the wheat plant grows poorly when it's too hot.[30]

But it's not just higher temperatures that are affecting agriculture. There are four more major warming-related culprits. The first is the increase in extreme rainfall, which raises the chance of crop failure and leads to the erosion of fertile soil. The second is sea-level rise, which causes the salination of groundwater and the flooding of arable land. The third culprit is an increase in the cases of drought and a greater chance of wildfires that threaten crops. Finally, higher CO_2 concentrations in the air reduce crops' ability to absorb nutrients from the soil, as plants use CO_2 to produce more sugar at the expense of said nutrients.[31] This is one of the reasons why we're already having to consume more vegetables than fifty years ago in order to receive the same amount of essential minerals such as zinc.[32] One and a half billion people suffer from nutrient deficiencies.[33]

These consequences of warming are particularly problematic when they all happen at once, as they did in Russia in 2010. That

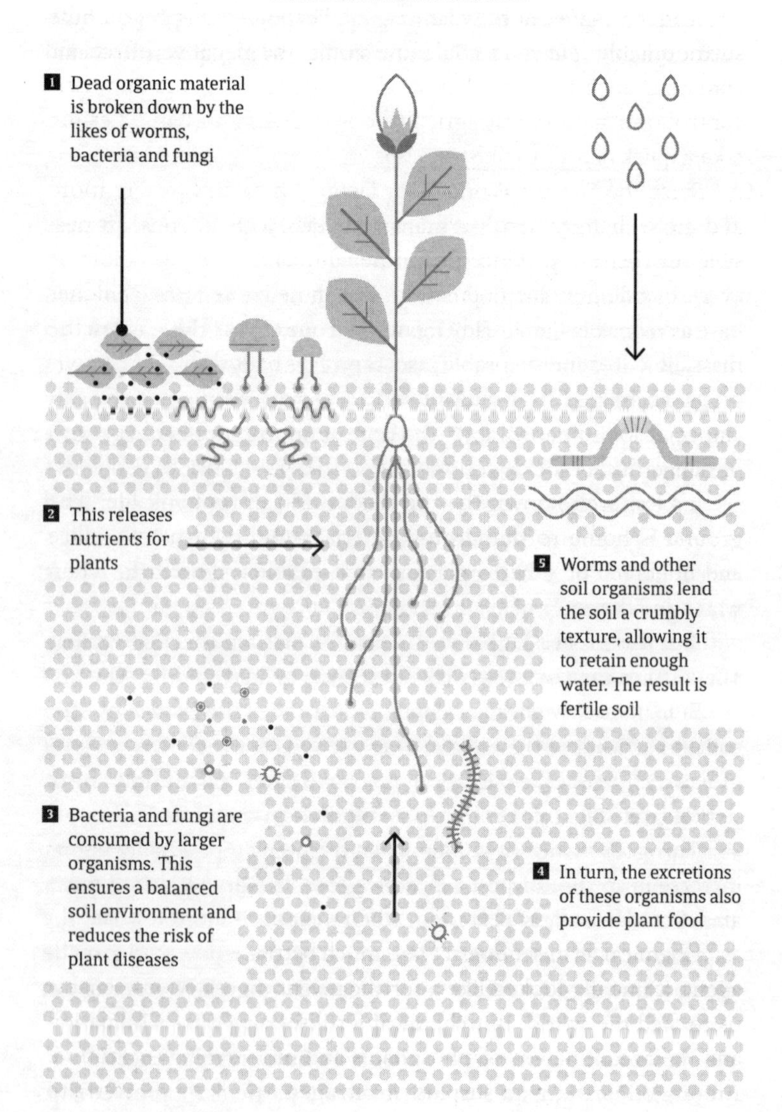
Our amazing soil life enables food production
Plants depend on the soil, which releases nutrients to them
1 Dead organic material is broken down by the likes of worms, bacteria and fungi
2 This releases nutrients for plants
3 Bacteria and fungi are consumed by larger organisms. This ensures a balanced soil environment and reduces the risk of plant diseases
4 In turn, the excretions of these organisms also provide plant food
5 Worms and other soil organisms lend the soil a crumbly texture, allowing it to retain enough water. The result is fertile soil

year drought, heatwaves and forest fires caused a third of the country's wheat harvest to fail. It led to a doubling of the wheat price on the world market.[34]

If global agriculture were in good shape, warming would be surmountable. Farmers could overcome the negative effects of climate change the way they've always done: by tightening their control over nature. But agriculture isn't exactly thriving. Let me take a quick detour to illustrate this.

Since the Second World War, farmers have been using more and more chemical fertilisers and pesticides. This has made it possible for them to produce phenomenal quantities of food for the ever-expanding global population. But these ingenious techniques have a downside. I've already mentioned one effect of pesticides: the mass die-off of indispensable insects such as bees that pollinate our crops. There are strong indications that some pesticides are also carcinogenic for humans,[35] and that their use contributes to the growth of dangerous, pesticide-resistant fungi that pose a risk to people.[36]

It's less well known that pesticides also disrupt soil life. The ground is home to numerous bacteria, fungi, worms, woodlice and hundreds of other organisms that enable the earth to retain water and the roots of plants to absorb nutrients. Life is impossible without this process: it promotes plant growth, purifies water and safely stores carbon from plant litter in the ground.

But the use of pesticides – and the deployment of heavy agricultural equipment – causes a sharp decline in these essential processes.[37] Reduced soil life compromises crops' natural immunity to pests, causing them to become even more vulnerable.[38]

This goes some way towards explaining why, despite a global increase in the use of pesticides, proportionally just as many crops are lost today as forty years ago, when their use started to soar.[39]

Chemical fertiliser doesn't fare much better. A powerful growth agent, it initially looked like a terrific invention that would enable us to feed the world. But chemical fertiliser also causes salination and the accumulation of phosphates and nitrogen in the ground. Among other things, a surplus of nitrogen causes soil bacteria to produce more of the powerful greenhouse gas nitrous oxide (N_2O), also known as laughing gas. This gas stays in the atmosphere for an average of 120 years and retains approximately 265 times as much heat as CO_2 does during this period. It's responsible for approximately 6 per cent of today's warming.[40]

*

Unfortunately, the problems caused by the excessive use of (chemical) fertiliser aren't confined to the production of laughing gas. Today's agriculture sector is responsible for 20 to 30 per cent of all greenhouse gas emissions, making it one of the main agents of global warming.[41]

This is because we're still cutting down trees to create fields, because our livestock emit the powerful greenhouse gas methane and because the agriculture sector itself is a major consumer of fossil fuels. We use oil and gas to drive tractors, to transport containers and planes full of food around the world and to dry crops. And to top it off, chemical companies use oil to manufacture pesticides and natural gas to produce chemical fertiliser.

In short, not only is the way in which we produce our food inextricably linked to global warming, but it will also take a direct hit as a result of it.

We're already coming up against the limits. Arable land that's tilled with machines, pesticides and fertiliser can't do without them after a while. Farmers can respond to this by using even more fertiliser and pesticides – expensive measures that are effective for a short time. Then the land starts to fail.

Globally, 20 per cent of fertile arable land is now seriously depleted.[42] That means that this land is no longer suitable for agriculture, or at the very least supports much less growth than before. The fact that the amount of fertile land is decreasing while there are ever more mouths to feed is a downright alarming development.[43]

Dangerous cocktails

I'm not writing all this to send you spiralling into depression, but to present you with a clear picture of the situation: how the earth we live on is changing, and how it's going to affect us. As you've seen, it's the secondary effects of global warming that will get us into trouble. Compare it to the corona-pandemic: it's probably not the virus itself which keeps you up at night. You might get sick, I hope you don't, but if you do you will probably survive it, just like you will survive an even higher CO_2 concentration. But it's the fact that intensive care beds might run out that should alarm you. It's that you can lose your livelihood as a result of the economy tanking – it's the poverty looming for half a billion people.[44] It's the food supply that could be severely disrupted by the virus, prompting the UN to warn of the worst hunger emergency ever.[45] The corona-crisis showed us how fragile we are for these kinds of cascading effects, and how climate calamity might hurt us.

Of course there's a lot we don't know yet. It's possible that we'll develop crops that are much more heat and drought resistant than we now think possible. But the chances of unpleasant surprises won't simply disappear, because the effects of warming and other man-made impacts on the environment interlock and form dangerous cocktails.

For a good example of such a cocktail I return to the sea. What I didn't mention before, but you might as well know by the end of this uplifting chapter, is that warmer seawater carries less oxygen. This is extremely unpleasant for fish, which need oxygen to breathe, and so large numbers of them are already moving away from waters that are warming up too rapidly for their taste.

When the increased heat of surface water combines with pollution from chemical fertiliser, the overall effect can be disastrous. That's another thing I haven't mentioned yet: excess fertiliser from agricultural land and cattle farms that's rinsed off fields by rain ends up in rivers and in the sea, and so contributes to the water's 'eutrophication', a fancy term for what's essentially the fertilisation of that water.

Algae, bacteria and some aquatic plants love this, especially at higher temperatures, which is why algal blooms are a common sight in ditches and canals during warm summers. Disadvantage: where algae proliferate, oxygen levels in the water drop even further, causing fish to suffocate and float to the surface – dead. A morbid summer spectacle that's been seen the world over in recent years.

Out at sea, the combination of increased heat and greater pollution from chemical fertiliser is also taking its toll.[46] In at least 400 separate locations the seawater no longer contains any oxygen at all during the summer.[47] Jellyfish, bacteria and other microbes are all that's alive in these 'dead zones'. One such zone in the Gulf of Mexico covers a surface of some 14,000 square kilometres in summer – about the size of Northern Ireland. The total area conquered by these dead zones is now four times as big as it was in 1950.[48]

The unforeseen consequence? When all the oxygen is gone, some bacteria thrive, among them those that produce methane, nitrous oxide, CO_2 and 'rotten egg gas' hydrogen sulphide.[49] The latter is poisonous, while the other three contribute to global warming. As I said: a dangerous cocktail.

*

There are more of these cocktails on the menu. The combined action of extra heat and too much fertiliser is only one example of a principle at work in many areas: one effect triggering another.

Look at what happened in Mozambique in March 2019. Cyclone Idai brought widespread devastation: hundreds of casualties, tens of thousands of buildings destroyed, hundreds of thousands of people left homeless and extreme rainfall that caused large-scale flooding and the destruction of more than half a million hectares of agricultural land.

This is exactly the sort of disruption that comes with a warmer climate – and in turn triggers other problems. Two weeks after the storm, the World Health Organization warned of a second disaster: more and more cases of cholera were being reported, and an increase in malaria was anticipated. Assistance to HIV patients had been disrupted.

It was a textbook example of a dangerous chain reaction on

an ever hotter earth. While it can't be *fully* attributed to global warming, it can't be seen in isolation from it either: climate change compounds the risk of such disasters.[50] In the aftermath of a catastrophe like this, the damage can be limited via emergency aid and vaccination campaigns, but humanitarian aid can never eliminate all of the suffering.

Three weeks after the disaster in Mozambique, 3 million people in the area affected, of whom 1.5 million were children, were in urgent need of healthcare, clean drinking water and sanitation.[51] Another three weeks later, the north of the country was hit by Cyclone Kenneth, one of the most powerful storms ever to make landfall on mainland Africa.

Months on, at least 75,000 people remained displaced, while 1.6 million relied on food aid for survival.[52]

Tipping points and the dangers of a hothouse earth

It's tempting to think that gradually reducing emissions is enough, that the climate will calm down if we do. That all's well that ends well.

But the problem is that there will come a point when a warmer earth starts following its own rules, not ours. A hotter planet can take over from us, can make itself even hotter.

(If you're starting to feel the urge to curl up in a foetal position in a corner of the room: I share your pain. It gets better after this chapter.)

The way global warming reinforces itself is plain to see in the Arctic. At least two thirds of the area's sea ice has disappeared in recent decades.[53] More and more of the white surface of the ice has been replaced by dark seawater, which reflects less sunlight and converts more of it into heat – just like a black shirt is warmer than a white shirt in the sun. It means that warming in the Arctic intensifies as more ice disappears. The consequence: the Arctic has warmed two to three times faster than the global average.[54]

The fact that warming can reinforce itself is something that the first climate scientists in the late nineteenth century already figured out. So the existing climate models take into account that a less white globe will warm up faster. But nobody knows exactly *how fast* the ice in the Arctic will disappear – estimates are all we have. And climate change can reinforce itself in many other ways, none of them easy to predict.

Look at the permafrost, for example, the permanently frozen ice on the tundra around the Arctic. It stores the greenhouse gas methane, together with plant litter and other organic matter that bacteria convert into CO_2 when the ice melts. The rule of thumb is: more heat equals more melted permafrost equals more greenhouse gas emissions.[55]

Above the Arctic Circle in Alaska, meltwater lakes have formed, from which CO_2 and methane are bubbling up.[56] Luckily, current emissions there are partially compensated for by new plants, which can grow where the ice disappears.[57] But there's a tipping point, after which the thawed permafrost will send more greenhouse gases into the air than the new plants are able to absorb. Passing that point will be disastrous, because the permafrost contains a total of 1.5 billion tonnes of carbon, more than twice as much as there currently is in the atmosphere.[58]

The problem is that nobody knows where the tipping point is.

*

Scientists do know that nearly all natural systems have such tipping points. A rainforest that's drying out because of heat, for example, will eventually turn into a savannah – which retains much less CO_2. And when a forest goes up in flames, it will release all the CO_2 it has stored for decades in one terrifying burst. In just four months at the end of 2019, Australia's blistering bushfires released as much carbon as the UK economy in one entire year. New trees can capture those emissions again, but it takes between ten and one hundred years before they reach their former size – and with it the same carbon storage capacity.

Picture yourself in a canoe. You can sway back and forth. You can tilt a little to one side and bounce back. But if you do it again and go a little too far this time, gravity will take over and your canoe will overturn. You'll go from one stable state – safe in your boat – to another: upside down in the water.

We won't be crossing these tipping points in natural systems overnight, and certainly not all at once. And some we may never cross. But the reason I dwell on them is that once we've crossed one there's no going back. Try as you might, but you can't put methane back into the permafrost. Unless we have another ice age, which at today's CO_2 concentration levels would be in 100,000 years at the earliest. And it's not as if we're waiting for that to happen.

Despite their irreversibility, tipping points play only a marginal role in the climate debate. The IPCC warns of 'large-scale irreversible events',[59] but because these processes are so hard to predict[60] they're not incorporated into the climate models on which our politicians base their policies.[61]

*

History teaches us that it's unwise to ignore the existence of tipping points. Climate scientists have reconstructed a specific period of rapid warming that shows this particularly well. It took place 56 million years ago.

There were no people at the time, and the earth was a boiling-hot place: more than 10 degrees warmer than it is today. Atmospheric CO_2 concentrations were over five times what they are now, there were no ice caps and the sea level was 100 metres higher. Rainforest covered the land masses around the Arctic, crocodiles swam the seas and hippo-like creatures walked the land.[62] Sounds weird, I know, but it gets worse, because it grew even warmer. It's thought that the trigger was a succession of major volcanic eruptions, causing a further increase in CO_2 levels in the atmosphere.[63]

And that's when it happened: a decisive tipping point was crossed. Reconstructions indicate that frozen methane in the seabed began to thaw. Countless molecules of the powerful greenhouse gas bubbled up to the surface, entered the air and caused an additional 5-degree warming.[64]

The crocodiles in the Arctic now had palm trees to lie under. The Arctic Ocean reached a temperature of approximately 23 degrees, while the seawater near the equator went up to 37 degrees. It began to rain harder, as you'd expect, since warmer air retains more water vapour. During the rainy season, rivers in what are now the Spanish Pyrenees shunted eight times more water than before the decisive shift.[65] Imagine the amount of water that has to pour from the sky to make a river eight times wider! To this day, large boulders in the mountainous landscape bear witness to the cloudbursts from that time.

In short, the earth became a totally different place. Over time, the slow-grinding wheels of mother nature did their job. But it took 200,000 years. Once all the excess methane and CO_2 had disappeared from the atmosphere, a colder period dawned again. New ice crystals containing the greenhouse gas methane formed at the bottom of the ocean.

And there they remain to this day: ready until the earth crosses another tipping point, once again releasing the gas in unfathomable quantities.

Whether and when that will be, nobody knows, but it's clear that we're trying our utmost to make it happen.[66] Should it come about, billions of people will witness what happens when you give nature a push, and another push, and another.

In that case we could, as a research group put it in 2018, end up on a 'Hothouse Earth': an earth that's 'uncontrollable and dangerous' and 'inhospitable to current human societies and to many other contemporary species'.[67]

I wish that what I've just told you were an eccentric little theory, a fable dreamt up by climate alarmists. But no. All bets are off. The history of our planet has known five extinction events, during which a huge percentage of all life on earth was eradicated. We all know the cause of one of those mass extinctions: the meteorite that killed off the dinosaurs. The other four didn't happen in a flash, but played out over a period of thousands of years.

They were caused by climate change.

PART II

WHERE ARE WE HEADED?

Chapter Four

IT'S NOT TOO LATE

When the American chemist Frank Sherwood Rowland discovered, one day in 1974, that there was a hole in the ozone layer, on returning from the lab, he told his wife, 'The work is going well, but it looks like the end of the world.'

For those who, like me, weren't born yet, let alone aware of all the fuss at the time, ozone is a gas in the atmosphere, one of the small group comprising 1 per cent. It not only retains heat, but also protects us against carcinogenic UV radiation from the sun.

Rowland found out that ozone is broken down by CFCs – refrigerants used in applications such as fridges and air conditioners, which can leak into the atmosphere when the machines end up on the rubbish heap.[1]

As soon as it was known that CFCs indirectly caused cancer, panic broke out. Newspapers printed article after article about the hole in the ozone layer. A significant proportion of the world's population might get cancer if the hole continued to grow. Some people, including Rowland, feared that it would lead to the end of the world.

Fortunately things turned out differently. Within two years of the discovery, the international community succeeded in drawing up legislation to ban the use of CFCs. The ozone layer has been recovering since then, and you hardly hear anyone talk about it anymore. The Montreal Protocol, in which countries agreed to ban CFCs, became known as the most successful international environmental treaty ever. It's still in effect.[2]

Climate change is a far more complex, all-encompassing and

unruly problem than the hole in the ozone layer. Nevertheless, right now, as in the 1980s, it's not too late. The greenhouse gases we have already emitted will continue to have an effect for centuries to come, but the shocking possibility of a hothouse earth, as I described above, is not yet a certainty.

With today's knowledge, further climate change is a choice, not a given. What happens from here on in is up to us.[3]

The difference between climate change and the apocalypse

The big question is how far we let it go. All countries of the world agreed in the first climate convention in 1992 that they wanted to prevent 'dangerous anthropogenic interference with the climate system'. It took until 2009 to come to a definition of what precisely constituted 'dangerous', but then the international community came up with a nice round number: a temperature rise of 2 degrees Celsius compared with the average world temperature in the period prior to 1750, when the Industrial Revolution began. Those two degrees were the hard border that the world couldn't be permitted to cross.

During the Paris climate conference in 2015, after long insistence from climate scientists, representatives of low-lying islands and environmental organisations, a second goal was added: a maximum of a 1.5-degree temperature rise.[4]

That half-degree difference between 1.5 and 2 degrees matters immensely. According to a recent IPCC report, at a 1.5-degree temperature rise it's likely that 14 per cent of the world's population will be exposed to extreme heatwaves every five years. At 2 degrees the figure affected in this way rises to 37 per cent.[5] The difference translates to 1.7 billion people (!) confronted with a very concrete question: whether they would still be able to work, play sports or cycle outside on hot summer's days without fainting, getting a headache or throwing up as a result of the heat.

Or worse. At a 2-degree rise, large metropolises where it's already warm, such as Kolkata and Karachi, will have to endure annual heatwaves which will cost part of the population their lives.[6]

People on low incomes, who can't afford air conditioning, will be hardest hit.

And so it goes on. At a 1.5-degree rise, 180 to 270 million people will be spared a life of constant clean water shortages.[7] At 1.5 degrees, some tropical coral reefs might still survive, whereas at 2 degrees they will all die out.[8] At 2 degrees, 18 per cent of insects will lose more than half of their habitats. At 1.5 degrees, 6 per cent of insects will have to deal with this devastation.[9]

And so on.

Climate change isn't a question of yes or no, but of *how bad*.

We determine how bad it gets. We don't have the luxury of checking out and saying: this isn't going anywhere. The speed with which we reduce emissions will determine the level of CO_2 concentrations in the atmosphere, and that determines how much the climate heats up.

*

We don't have *complete* control over the future. The chance of passing climate tipping points becomes ever greater once we go beyond 2 degrees. But even if we pass such a tipping point, for example, when more greenhouse gases leak out of melting permafrost than plants can reabsorb, the world won't *suddenly* change unrecognisably, coastal areas won't *immediately* be submerged, people in areas of drought won't *instantaneously* pack up and try their luck elsewhere.

Global warming may cause large-scale devastation locally and regionally, but most of the consequences will manifest themselves gradually, over a period of months to years. There will be no Judgement Day, nor will the earth burst into a sea of fire, even if the stock photos accompanying news stories about global warming persist in suggesting as much.

The American media in particular have a habit of portraying global warming as the end of the world, with bombastic headlines such as 'Apocalypse Soon: 9 Terrifying Signs of Environmental Doom and Gloom'. At the end of 2018 the rising star of left-wing America, Alexandria Ocasio-Cortez, literally said, 'The world is going to end in 12 years if we don't address climate change.' Others, like the activist group Extinction Rebellion's founder, Roger Hallam, have fantasised that 'the human race will be extinct in a decade'.[10]

That is utter nonsense. Between 2030 and 2050 we'll probably cross the threshold of 1.5 degrees of warming if we don't radically change course before that.[11] And what happens when we pass 1.5 degrees? Then it'll get even warmer and all the consequences of global warming will be more severe and dangerous. But that still doesn't make climate change some kind of Armageddon, nor a competition. There's no referee to whistle when the 'game' is over. Activists and politicians can keep on saying it, but there's no 'five minutes to midnight', because the clock will never strike twelve.

Well yes, eventually the doomsayers will be right: in a billion years the sun will become so hot that the oceans boil. That will be the definitive end of the story of humankind and the blue planet.

But it'll be a while yet. The process will take a good three thousand times as long as the entire history of *Homo sapiens* so far, to be precise. The successors of *Game of Thrones* writer George R. R. Martin could write another 300 million series of the show and in every one of those series the apocalypse could be announced without coming to pass. 'Winter is coming,' for the next 2.7 billion episodes.[12]

Meanwhile, billions of people will have to live with the consequences of climate change. Personally, I consider that the most important argument for limiting further damage. Every half – even every tenth – of a degree of global warming avoided is a plus. Ultimately, for millions of people, it's the difference between being driven from their homes by an increasingly unliveable climate or not. And I'm not talking about people in some far-off future either. It has already begun, even if you won't hear about it on the evening news.[13]

In any case, it's still far too early for the grim conclusion that climate change will lead to our demise. We still have every opportunity to prevent worse consequences and even, despite the rising mercury, to make a much better world possible.

What is needed for that? For a start, we can try, really try, to achieve the climate goals agreed in Paris. In order to keep within reach of the first goal – 1.5 degrees maximum – the IPCC says global CO_2 emissions would have to have dropped by 40 to 60 per cent by 2030 compared with 2010. So effectively halved in twenty years.[14] By 2050, emissions would have to be zero to limit global warming to 1.5 degrees.[15]

Failing that, we have to achieve zero CO_2 emissions around 2070 to keep it under 2 degrees.[16] Emissions of other greenhouse gases, such as methane and nitrous oxide, will also have to drop rapidly this century.[17]

To give you a sense of scale here: as a result of corona-lockdown measures, global emissions in 2020 were expected to fall by 8 per cent, a drop 'twice as large as the combined total of all previous reductions since the end of World War II', according to the International Energy Agency.[18] But to keep 1.5 degrees within reach, we'd need to realize a drop like that *every year* for the next decade – using humane methods instead of deadly viruses.[19] There's a challenge for you.

An important disclaimer in these figures is that we don't precisely know to what extent global warming will be self-reinforcing as the mercury continues to rise. To use the scientific jargon, researchers can't predict the precise 'climate sensitivity'. For instance, if more sea ice disappears at the North Pole than the standard models currently foresee, global warming in that region will also be self-reinforcing – because water absorbs more warmth from the sun than ice does.

It's impossible to completely eliminate these kinds of uncertainties in the models. As I described previously, some self-reinforcing processes – such as the melting of permafrost – cannot (yet) be modelled well. It may therefore turn out that we do everything necessary according to the current calculations to limit global warming to 1.5 degrees, only to pass that threshold *anyway* because global warming is uncontrollably self-reinforcing.[20]

The only way to be certain, it seems from all the IPCC reports,[21] is to reduce greenhouse gas emissions as quickly as possible.

Sadly that's not going all that well so far. Emissions are still rising almost every year, and we're heading towards a 3 to 4 degree rise by the year 2100. Nevertheless, in principle it's still *possible* to achieve the climate goals. In the words of UNEP, the United Nations Environment Programme, 'Now more than ever, unprecedented and urgent action is required by all nations.'[22]

In short, we need to stop all emissions of greenhouse gases in the next thirty years. That requires an unparalleled effort and we may very well fail. The way things are currently looking, the chances of us overshooting the 1.5- and 2-degree thresholds are horribly real.[23]

If it came to it, we could try to go no further than 2.5 degrees, and then 3 degrees, and then 4.[24] Sceptics on the sidelines will say we're always shifting the goalposts, and it's true – that's what you do if you don't want to slide towards a hothouse earth.

The negative human self-image standing in the way of action

That brings me to a question that has preoccupied me for years. Because if we really can still take action, why does it often not *feel* that way? Why can we imagine the end of the world so much more easily than a big turnaround?[25]

One answer is that we're much better at fantasising about the end of the world than about the alternatives. The story of the great flood, or another disaster to end civilisation, is the oldest story in the world. Every culture in history has a story like this, and on Netflix you can indulge in apocalypse stories night after night. (My personal favourite is the millennium bug. Have a quick google if you were born post-2000.)

Under this cultural baggage is a deeper explanation, a stubborn idea about who we are and what we're capable of. The most important reason for our meagre trust in an effective way of tackling the climate problem, I think, is that many of us have begun to believe, deep down, that humankind itself is the problem.

I'm talking about the negative human self-image that is now so widespread that we no longer notice it: the idea that we're inherently selfish and therefore will never work together for the common good. That we're doomed to fail, that we deserve it.

An example. In August 2018, *The New York Times Magazine* published an article about the rise of climate science between 1979 and 1989.[26] According to the writer, Nathaniel Rich, that period was the decade in which we 'had an excellent opportunity to solve the climate crisis'. There was already considerable consensus as to the dangers of climate change; political discord remained limited. 'The conditions for success could not have been more favourable,' Rich writes. All the facts were known and 'Almost nothing stood in our way – nothing except ourselves.'

Humans simply don't have it in them, Rich writes: 'Human beings, whether in global organizations, democracies, industries, political parties or as individuals, are incapable of sacrificing present convenience to forestall a penalty imposed on future generations.'

If that is correct, we might as well give up. According to this view we'll never lift a finger to stop global warming because we're only living for ourselves. 'Our actions have demonstrated,' Rich says, that the lives of our grandchildren 'mean nothing to us'.[27]

Could that be the case? Is it really so bad?

In my view it's this very image of humanity that stands in the way of collective action. All the more reason to stop and think about it a bit longer. The origin of the idea that humans are egotistical by nature takes me back to the Enlightenment, to the time of Descartes.

*

One of the most important Enlightenment thinkers, British philosopher Thomas Hobbes, was among the first to put into words what many people already believed at the time: that humans were primarily driven by a desire for personal gain. If we had no states and armies, a human life according to Hobbes would be 'nasty, brutish and short'. Civilisation, he suggested, is a thin veneer restraining our worst impulses. As soon as this protective layer is taken away, we'll beat one another's brains in.

So it's not all that surprising that our endless longing for prosperity has come at the cost of countless animals and natural habitats. It confirms what we already believed: that deep inside we're brutish and egotistical.

In recent centuries this idea gained the status of a biological fact: people were condemned to competition and egoism. Darwin's evolutionary theory played an important role here. The phrase everyone associates with Darwin today, 'survival of the fittest', came to be known as a merciless law of nature: only the strongest survive. We became who we are through endless conflict and 'natural selection' of the strongest. In 1976 the evolutionary biologist Richard Dawkins claimed in his bestselling book *The Selfish Gene* that even our genes express this trait. Egotism is embedded in our DNA.

This narrow interpretation of the theory of evolution gained momentum in the early twentieth century, when increasing numbers of economists began to claim that 'the invisible hand of the market' would be best placed to do its work if every individual and every organisation were to strive exclusively for their own interests.[28] As in the case of our genes, the strongest and best would float to the top, and those winners would then push humanity forward on the long march of progress. Optimal market forces, low prices and robust growth figures would be our reward. And this was not only the best way to set up the economy – it was the *only* way, as it reflected our deepest being.

Rather depressing ideas. But how reasonable are they genuinely? Do we have a fatal flaw embedded deep within us? How egotistical are we really?

When it comes to those 'selfish' genes, Dawkins later admitted his book would have been better titled *The Cooperative Gene*.[29] He hadn't taken much notice in the 1970s, yet throughout the past century evidence had been piling up. In nature, collaboration is at least as important as conflict.

In 1967, Lynn Margulis demonstrated that creatures that live together intensively become co-dependent. The engines of our cells, for example, known as mitochondria, were originally independent bacteria, but at some point they were absorbed and our cells acquired their own power supply.

Human DNA is full of these kinds of traces of viruses and bacteria. Right inside our cells and deep inside our genetic code, we're the product of collaboration – and the same goes for other animals. Every fibre, every one of our cells is interwoven with other life.[30]

Darwin in fact already suspected this: he was inspired by the 'web of life' that Alexander von Humboldt talked about, and he believed that natural selection would steer things towards *more collaboration*, because groups of people who work together would do better in evolutionary terms than groups with many egoists.[31] Of course he was right: you only need open your eyes to those around you to see that collaboration is an evolved trait in humans, and advantageous to most people.[32]

This doesn't change the fact that competition is an important mechanism in evolution. Or that people can be selfish. Both are also true – and, in our current economy, selfishness is widespread. But it's complete nonsense to act as if it's our biological destiny to be *exclusively* selfish. The idea of 'fitness' in 'survival of the fittest' applies just as much to being good at collaboration.

So yes, during the corona-pandemic lockdowns, some people broke the rules. Rebelliously they went out, they shook hands and sneezed where they pleased. But most of us decided to temporarily sacrifice our freedom so that others wouldn't get sick. It was a monumental collaboration: the self-curtailment of whole societies – of the kind you wouldn't observe in, say, a swarm of locusts.

*

People aren't just team workers; we also continually look out for one another. If someone falls off her bike, others are immediately there to help. In all kinds of emergencies – if someone is about to drown, if a fight breaks out – bystanders actually rush in to help, often without thinking.[33] The image we have of people standing in a circle watching an incident turns out to be wrong: generally someone intervenes.

Precisely at the moments you'd least expect it, our best qualities come to the fore. Time and again, research shows, when disaster strikes, we're there for each other.

We might be inclined to think that people in emergencies turn out to be the animals Hobbes thought us: egoists, looters, rioters. And sometimes riots do break out, sometimes shops are plundered. Of course there are opportunists. But most people, even in times of calamity, have no malice in them.

What would you do if the city where you lived was underwater and you (or your child) were hungry? Would you break a supermarket window if you had no other options? Desperation isn't the same as malice.

'In fact,' writes American essayist Rebecca Solnit on the basis of her research into human behaviour in times of disaster, people in such situations are 'calm, resourceful, altruistic, and creative'.[34] They search together for ways of coping until the emergency services arrive, and often succeed remarkably well. New communities come into being, and people go to great efforts to help themselves and others.[35]

It is in those moments, writes Solnit, that it seems as if a starved part of ourselves is nourished,[36] a part that we appear to keep on forgetting.[37]

Sure enough, when Covid-19 began killing people around the globe, we leapt into action. In the UK, the Netherlands and elsewhere, doctors and engineers teamed up to produce face shields and ventilators – repurposing, among other things, snorkelling masks. In China, when public transport was suspended, volunteer drivers taxied medical workers to the hospital and back home. In India and South Africa, communities self-organised the distribution of aid packages with sanitiser, bottled water and food for those most in need. Everywhere, young people started doing groceries for people with underlying medical conditions and for the elderly, as well as picking up their prescriptions.[38]

People also hoarded toilet paper, it's true. But most of what we did in response to the disaster was solidary and kind. People sang songs for one another from balconies and rooftops – someone even invented 'balcony bingo' (I'll let you guess how that works). We applauded healthcare workers and called our mums. It was as if this crisis revealed our true selves. We want to be there for each other.

A new story

So the idea of a climate apocalypse is misleading, and the negative human self-image considered 'normal' is in fact extremely one-sided. Phew! We still have a future.

And we still have a good many options to shape that future for ourselves. The fact that the story of 'humans lording it over nature' has been dominant so far doesn't mean that it has to stay that way. Nor do we have to keep on telling each other that we 'just don't have it in us'. That people are only focused on the short term, that we're selfish, that we always want profit – they're all stories, fairly recent inventions and now widely accepted. That creates room for manoeuvre, because it means that they can be replaced with different stories – that, in fact, is how we shape ourselves.

What we eat is a perfect example. As I write this, most people in the West consider it perfectly acceptable to eat animals: pigs, cows, chickens. But dogs? We don't eat them here. Whereas in South Korea that's completely normal.

What makes the difference? There are no scientific arguments for eating pigs and not dogs. Both are approximately as intelligent as each other. Nevertheless most Westerners would consider it completely barbaric if I were to eat my dog Moos.

The only explanation is the story we tell each other about these animals. Dogs are 'our best friends', pigs produce delicious bacon. 'Stories give rise to rules,' the American writer Jonathan Safran Foer concludes in his book *Eating Animals*. Bacon yes, dachshund no.

Because stories determine what we do and don't consider normal, they also affect the way we behave. 'We are made of stories,' writes Foer.

*

Which story do we tell each other about climate change? Do we tell each other that it's too late and that the destructive impulse is in our genes? Or do we tell a new story, based on a more realistic view of humankind, a story of recovery?

Whichever story we choose from here on in, we'll be presented with the results of our actions, and we'll try to adapt to them as best we can. The question is *how* we do that, and especially, how much damage we can still manage to prevent.

In order to illustrate the choices that stand before us, I'll sketch two contrasting future scenarios. These are not *predictions* – I'm not casting myself as a prophet. They're sketches of *possible* futures, intended to show that there really is a choice to be made.

In both cases I begin with a description of an imaginary world in the year 2050 – a world in which many of us really could end up – and proceed to show, based on current and past developments, how and why we could get to that point.

Chapter Five

FUTURE SCENARIO 1: WALLS

Year: 2050
Somewhere in Western Europe
Global warming since the Industrial Revolution: 2.6 degrees Celsius

It has become difficult to exercise outdoors, as the summers are too hot these days. For a while it looked as if we were going to have the sort of weather that the Mediterranean used to enjoy, but we soon overshot those pleasant temperatures. This summer we had dozens of long, languid days with the kind of stifling heat that gives you headaches. One of the targets set by the Ministry of Climate Control is that by 2055 all households should have air-conditioning, but everybody knows there's not enough money to realise this ambition.

The economy is struggling. Our export of agricultural products has declined due to heat and drought, increased soil salinity and torrential rainfall. Food prices have rocketed worldwide, while the costs of coastal protection are steadily mounting.

Dyke elevations are keeping the inlands dry for the time being, but the overriding concern now is that more and more land will flood anyway during the rainy season, when lengthy downpours make rivers burst their banks.

The National Flood Plan provides for the sacrifice of nature reserves, farmland and small villages. But despite these drastic measures, the government can't guarantee that residents of major cities will keep their feet dry. Only the providers of climate-proof housing projects are in a position to offer such assurances, but you have to be a multimillionaire to live in one of those.

Fortunately we're shielded from the unrest elsewhere in the world, as Europe's Ministers of Defence have pledged to protect the continent's border walls. But that doesn't stop us from seeing footage of people trying to climb over, in search of a better life.

The picture is bleak, and it's hard to remain optimistic about the future now that the tundra is emitting more and more greenhouse gases, nearly all tropical coral reefs have been destroyed and scientists are saying that the Amazon rainforest is rapidly drying out and beyond saving.

*

As you can see, climate change has continued to escalate in this future scenario. Governments are doing all they can to manage the consequences, because we have collectively failed to sufficiently mitigate global warming. A safe and comfortable life is set to become the preserve of a wealthy minority, even more so than today. In this scenario, society is in a state of constant heightened tension, which could spiral even further out of control at any moment.

How would we end up in this future world?

The astonishing thing is that not much needs to change for this to happen. If we allow the current financial and political systems to continue as they are, we'll automatically end up in this situation. Drawing on developments past and present, I will demonstrate why this is a realistic scenario.

Business as usual

Let's take a look at Shell, one of the biggest oil companies in the world. In recent years I have done a great deal of research into the Anglo-Dutch firm. I interviewed dozens of employees and pored over internal documents from the 1980s and 1990s that were leaked to me and that reveal what the company knew and what it was thinking about the climate. The conversations and documents all point to one conclusion: Shell was extremely worried then, and remains so now.

Here's why. In an internal study carried out in 1986 and

published in 1988, Shell researchers warned of 'relatively fast and dramatic changes' to the earth's climate, which would impact 'the human environment, future living standards and food supplies, and could have major social, economic and political consequences'.[1]

In 1991 – fifteen years before *An Inconvenient Truth*, the documentary about former US vice president Al Gore's efforts to raise public awareness of global warming – Shell made a film warning about the consequences of climate change. 'Action now is seen as the only safe insurance,' the voice-over says.[2] With this film, aimed at the general public and shown at special screenings, particularly in schools and universities, Shell wanted to initiate a public debate about possible solutions.[3]

Not all oil companies were as forthcoming as Shell about this, but they'd certainly all long been aware that there was a problem.[4] As early as 1968, the whole American oil industry was warned that CO_2 emissions were a leading cause of climate change. It was their own trade association, the American Petroleum Institute (API), that sounded the alarm in an internal document.[5]

How is it possible, then, that Shell – like the other oil majors – continues to extract fossil fuel, while also investing heavily in exploration for future drilling?

The answer is simple: the companies have no choice. There is demand for energy, shareholders expect dividends and competitors will gobble up any bit of market share that Shell surrenders. If Shell doesn't extract the oil, so the story goes within the firm, another player will.[6]

Every Shell worker is a tiny cog in this big machine, an employee told me,[7] and changing direction is something not even the chief executive can do. Leaving fossil fuels in the ground is 'not the mission shareholders put me on this earth for', Shell CEO Ben van Beurden said in 2016.

*

Shell is an example of a company full of people with the best intentions who are nevertheless incapable of achieving what's necessary:

transformative change.[8] The company is pressing ahead with what it has always done, even though the extraction of new fossil fuel is becoming harder and more expensive and, like many other fossil-energy-based companies, it has incurred debts in recent years to carry on paying dividends.

In May 2016, I was at Shell's annual shareholders' meeting, a recurring ritual in the seaside resort of Scheveningen, when an investor asked the board whether the company ought to perhaps reduce its dividend since it was eating into its balance sheet to pay out.

The then CFO, Simon Henry, an amiable Brit, said he wished things were different, but that the majority of shareholders simply wanted to see maximum returns. In other words: Shell's board finds itself between a rock and a hard place. It took the corona-crisis to finally make Shell cut its payout by two-thirds amid serious concerns over its long-term future.

Other private oil and gas companies are in a similar bind. It's all about the short term: the next quarterly report, the latest profit figures. In this context gradual change is the best we can hope for. At present, on average, only 3 per cent of all global investments made by the listed oil companies is spent on clean energy (and on expensive advertising campaigns about this clean energy).

Oil companies earn a lot more from a barrel of oil than from a wind turbine, which is why they're still putting far more money into finding new reserves and coming up with new ways of sucking oil and gas out of the ground. Shell excels at this, using algorithms to make drilling more efficient.[9]

To safeguard the profitability of their investments, oil companies are forced to go all out to defend their interests – for instance by using an army of lobbyists to dilute undesirable climate measures in London, Brussels, Washington and Beijing. Since the Paris treaty was ratified, the big listed oil firms have spent over $1 billion worldwide on their 'climate lobby'.[10]

You can think of this what you will, but the fact remains that these companies feel this is their only option. 'The earth is dying,' Shell non-executive chairman of the board Charles Holliday acknowledged in October 2019. Yet Shell has 'no choice but to invest in oil,' CEO Ben van Beurden said in the same week.[11] Shell is still in the bind it identified in the 1990s: how can we protect both the economy and the climate?

'We all suffer if economic activity degrades the environment,' Shell CEO Sir Mark Moody-Stuart said during a speech in Jakarta in 1994. 'Yet clearly, our own and future generations can only meet their needs through economic development.'

Governments are forced to maintain the same balancing act between growth and green. Like the fossil fuel companies, they realised a long time ago that something was seriously amiss. As early as the 1960s, widely debated international research reports began appearing, addressing the 'limits to growth'.

It was Margaret Thatcher who noted, in a speech to the Royal Society in 1988, that 'the health of the economy and the health of our environment are totally dependent upon each other ... Protecting [the] balance of nature is therefore one of the great challenges of the late Twentieth Century'.[12]

But here's the irony: in the same period, shortly after the fall of the Berlin Wall, capitalism well and truly conquered the world. The notion that 'the market' knows better than 'the government' prevailed. It was thought that unleashing the free market in as many sectors as possible would improve services and bring down prices. Just when the environment demanded state intervention, governments relaxed their controls so the market could work its magic.[13]

In reality, governments didn't disappear – from any market. 'Free markets' can only function when states determine the rules of the game with things like laws, tax credits and trade agreements. And so governments remained closely intertwined with the development of the economy, and in particular with the extraction of fossil fuels. They leased the land and permitted the drilling. Often, they helped experiment with new techniques. Take the recent rise of hydraulic fracturing or 'fracking' in the US – a controversial drilling technique to extract natural gas and oil from shale rock. It was made possible by more than three decades of government support and subsidies from the Department of Energy.[14]

Through investments in and levies on the extraction of raw materials, the government budgets of many countries have come to depend on oil and gas extraction. While not every nation is as dependent on oil for its income as Kuwait, Iran or Saudi Arabia are,[15] no economy on earth can function without oil, gas and coal. Those that don't produce these fuels themselves import them – or the products manufactured out of them, such as plastic and fertiliser.

So the economy of just about every country is directly implicated in the ongoing depletion of the earth.[16] The world economy has entered into a pact with fossil fuels, and whoever gets it into their head to try to break that pact will also break their economy.[17] Altogether, over the past two centuries, national governments and fossil fuel companies have invested some $25 trillion in mines and boreholes, pipelines and oil tankers, refineries and power stations.[18]

To put things into perspective: that amount could fund UK healthcare for over 100 years. But nobody can walk away from these investments. And they're only increasing – every year sees an additional $1 trillion worth of fossil fuel infrastructure.[19] China alone is overseeing a total investment of $12 trillion up to 2030 along its humongous 'Belt and Road' infrastructure project that spans 126 countries. And as things stand, most of that money will be spent on good old-fashioned fossil fuel development.[20]

At least 40 per cent of all planned investments in oil, gas and coal up until 2025 would have to be scrapped in order to meet the Paris climate targets and to keep below 2 degrees of warming.[21] The corona-crisis is likely to shrink fossil investments in the years to come, but if our travel and road transport recover from lockdown, so could the demand for oil and gas.

National governments have a big say in all this. The world's twenty richest nations alone – the G20 – subsidise the exploration of new fossil reserves to the tune of nearly $90 billion a year.[22] Add to this official measures to keep down the consumer prices of oil, gas and coal and you're looking at state support worth $400 billion in 2018.[23] That's more than double the subsidies allotted to renewables.[24]

The US, for instance, is subsidising fossil fuel production at a rate of $20 billion annually.[25] In Germany, the use of oil in the transport sector is heaviliy subsidised by tax reliefs on diesel (for cars) and kerosene (aeroplanes). Energy-intensive industries in that country enjoy tax breaks on their energy bills worth almost €10 billion a year.[26] In the UK, the main beneficiaries are the providers of passenger transport services (thanks to a zero-rate VAT), households (reduced VAT on fuel and power consumption) and oil and gas companies (fiscal support worth £665 million a year).[27]

Similar patterns can be found in the agricultural sector. Europe spends €65 billion annually on agriculture. Part of that money (around 30 per cent) is formally earmarked for 'greening', but it is 'not yet environmentally effective', one audit found.[28] It essentially remains income support that largely benefits big farm holdings. The more land, the bigger the subsidy – sustainability be damned.[29]

*

And so the system is self-perpetuating. States spend tens of billions a year on consolidating the bind we're in. With each new investment in oil, gas and coal it becomes a little less likely that we'll meet our climate targets. In response to the corona-crisis, some countries are making it worse by bailing out fossil-fuelled sectors like aviation without attaching any serious preconditions about sustainability. The whole idea that some sectors *ought* to be in decline – not rescued – seems lost on them.

That is painful because it means taxpayer money is spent rescuing big polluters who, by definition, trample the interests of taxpayers. And it's painful because we have plenty of options at our disposal for combating emissions: ending fossil subsidies is one, introducing high levies on CO_2 is another. But if a country is alone in introducing such measures, it would be at the expense of its economic growth, employment and the competitive edge of its industry. And nobody can afford that – least of all heads of government seeking re-election. So for individual countries and fossil fuel companies alike, small steps are the best we can hope to achieve.

If countries want to do *more* for the environment, they will have to come to an agreement together – for instance on a European level. But everybody is out to defend their own interests: Germany with its large automobile industry is blocking strict emissions standards for cars,[30] Poland with its many coal-fired power stations is grudgingly accepting Europe's climate policy, while insisting on exceptions for its own coal plants.[31] Just about every sensible measure is opposed by someone, and so the best possible outcome in Europe is a compromise. The country whose interests stand to be damaged the most by a specific planned policy determines the speed of change, and that speed is *slow*.

The world wasn't designed like this by some evil genius. It's nobody's intention to keep feeding a system we can no longer tame. But that doesn't alter the fact that we've reached complete deadlock.

Desperate diseases call for desperate remedies

It goes without saying that most politicians and fossil fuel companies want to do at least *something*. Trapped in an economic straitjacket, they come up with solutions that are well-intentioned, but often turn out to be half-measures. Plans that are both affordable and feasible, but that have no impact whatsoever – or perhaps even exacerbate the situation. Let me give you three examples of this type of policy, that show that half-measures only make matters worse.

The magic word in all three 'solutions' is *biofuel*. This is the catch-all term for agricultural waste, wood chips, leftovers from the food industry and fast-growing crops such as sugar cane, maize and palm oil. What all of these different 'fuels' have in common is that in their previous life they absorbed CO_2. Biofuels keep growing back and keep absorbing CO_2, making them look like the perfect replacements of oil, gas and coal. They recycle themselves, as it were. Unlike fossil fuels: when we burn those, we always add *extra* CO_2, previously stored as carbon underground, to the atmosphere.

Biofuels sound ideal, until you look at the bigger picture.

Half-measure 1. At the start of this century, many countries and energy companies became enthralled by the idea of mixing biodiesel and biopetrol into normal fuels. It meant that people could drive 'green' without having to get rid of their cars. It was what's known in political jargon as a 'cost-effective' solution for the consumer. It was perfect, at least on paper.

Now we know better. One of the most widely used raw materials for these 'clean' fuels is palm oil. Indonesia and Malaysia together supply 90 per cent of the world's palm oil. Primeval forest is cut down and peat land burned to create the necessary space for all the plantations. This means that the carbon that had been retained by the trees and the soil is released into the air as CO_2. Add to this the fact that chemical fertiliser, toxins and tractors are used to grow the palm oil and the true emissions of 'biodiesel' turn out to be up to four times *higher* than regular diesel.[32]

But while the EU may have pledged to stop importing palm oil for fuel as of 2030, the clearing of forest continues apace, and other, non-EU countries aren't planning to stop using what is on paper a 'green' fuel. Meanwhile the palm oil plantations are in direct competition with food production and demands for consumer drinking water in both Indonesia and Malaysia.

Half-measure 2. In the past ten years, more and more electricity producers have started using compressed wood pellets. Coal-fired power stations can either burn this biofuel alongside coal, or a power station can be refitted to burn only wood. The UK generates about a third of all its 'green energy' with wood pellets. It's low investment coupled with a direct return and no fuss. Sounds clever, right?

Under certain circumstances it is indeed a smart choice. If we convert sawdust from the paper industry or inedible stalks from the food industry into energy, we do the climate a favour. The alternative would be burning this waste directly without generating any energy (but still emitting particulates and CO_2) or dumping it, so it rots and ends up releasing both CO_2 and methane. This would be more harmful than using it for the production of energy.

But more often than not it doesn't work this way. Because there's not nearly enough wood, plant and food waste worldwide to meet the demands of electricity companies, forests are being cleared and – in the US and Canada in particular – special production forests used to produce sufficient quantities of wood pellets. The latest figures show that between 2008 and 2016 the import of wood pellets into Europe quintupled – resulting in plenty of emissions along the way. In the years to come, China, India, Japan, South Korea and Brazil are planning to step up the use of wood as an energy source.

And although it's true that new trees will eventually capture the CO_2 from power-station chimneys again, it will take between ten and a hundred years before we come full circle (most trees grow slowly). In the short term, power plants where wood pellets are co-burned actually emit more CO_2 than those that only burn gas or coal.[33] Therefore, like biodiesel, this 'solution' is often a step back compared to traditional methods.[34] Impressive! But that's not all.

Half-measure 3. The burning of compressed wood pellets also forms the basis for plans to achieve *negative* emissions later this century. The underlying assumption is that in the future we will be removing more CO_2 from the air than we emit. This is a textbook example of avoiding painful action in the short term in the hope of a magical technology saving us in the longer term.

Let's start by looking at the thinking behind this. Governments and companies that are sold on this half-hearted solution are expecting to reverse today's emissions if, at some point in the future, we can remove the main greenhouse gas CO_2 from the air and store it underground. It's akin to smoking strong roll-ups today because you're expecting a future inventor to come up with a way to suck the tar out of your lungs.

In climate terms this means that we'll continue our intensive use of oil, gas and coal for another hundred years. From 2030 until well into the twenty-second century, we'll be scrubbing the excess CO_2 in the atmosphere from the air. By bringing our net carbon emissions to zero later this century, we'll keep the warming below 2 degrees. Mopping up while the taps are still flowing – that's the 'solution' here.

How, you ask? On paper the mechanism looks deceptively simple: produce wood or reed crops such as willow, eucalyptus and elephant grass on a large scale. Burn these biofuels in a power station and capture the CO_2 that comes out of the chimney. It's a way of generating energy without emitting carbon. Pump the captured CO_2 underground, into an empty gas field, for instance. Do this year in, year out, on a large scale.

The result is a net drop in greenhouse gas in the atmosphere, because the CO_2 stored by the growing crops now ends up in underground storage. The idea is that this compensates for the emissions resulting from the use of fossil fuels. We can carry on using oil, gas and coal as usual, while simultaneously relying on this scrubbing technology.

This is not some marginal little plan. Climate scientists foresaw years ago that the world economy would still be running on coal, oil and gas until well into the twenty-first century. And so they sought a theoretical way of reversing the dangerous CO_2 surplus we'd be building up in the next century.

They came up with the technology I just described: the use of biofuels in combination with the underground storage of CO_2. Many governments loved it, so now the promised 'negative emissions' are totally ingrained in mainstream policy plans. So far the US, Canada, Norway, the UK and the Netherlands have shown the most interest in realising these ideas. Oil companies such as Shell also like the sound of them, because later this century they may be able to earn money by cleaning up the carbon they now extract from the ground.

The greater our dependence on fossil fuels in the decades to come – that's to say, the longer we continue emitting at our current level – the more dependent we will become on this 'solution' over time if we want to remain below 2 degrees. For the solution to work we would have to start actively removing CO_2 from the atmosphere from 2030 onwards. In the worst case it means that by 2050 we would have to compensate for a fifth of today's emissions by using biofuels in combination with CO_2 storage. That's almost as much as all the oceans together are now absorbing – a mammoth task.

The great appeal of emission compensation

Options for keeping warming below 1.5 degrees Celsius

— Net CO_2 emissions
▒ CO_2 emissions from fossil fuels and industry
░ Emission compensation through changes in agriculture and reforestation
▨ Emission compensation through scrubbing technology: biofuels in combination with the underground storage of CO_2

Option 1

Extremely rapid drop in emissions, no scrubbing technology

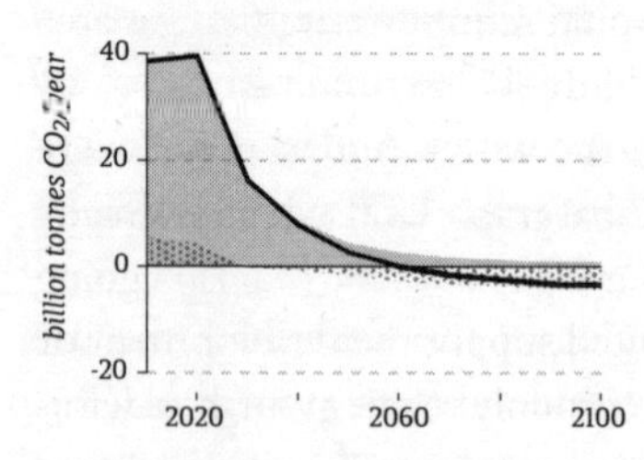

Option 2

Transition to sustainability, some compensation for CO_2 surplus via reforestation and scrubbing technology

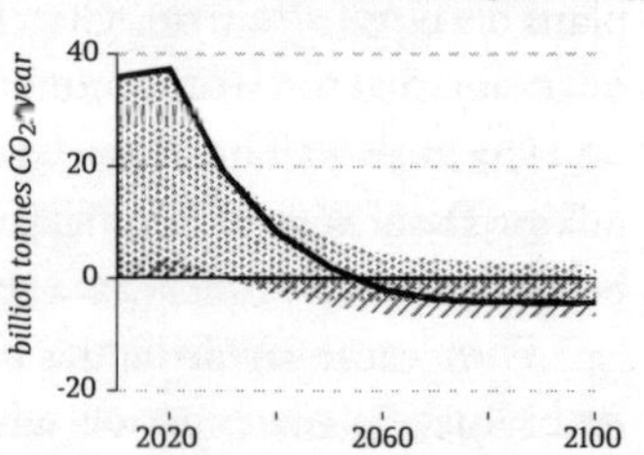

Option 3

Gradual drop in emissions, followed by lots of compensation via heavy use of scrubbing technology

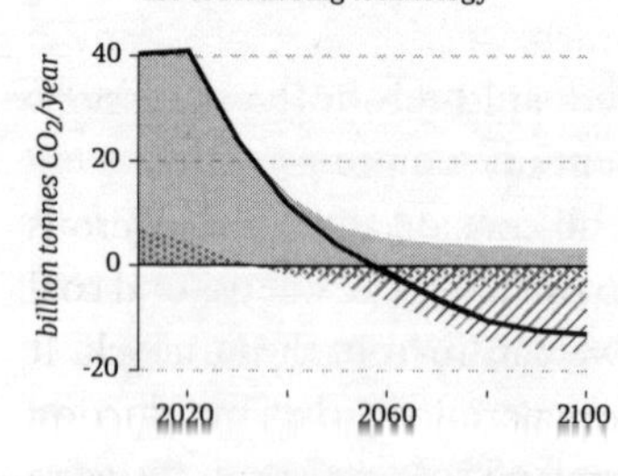

Option 4

Business as usual, followed by emergency measure: mass use of scrubbing technology

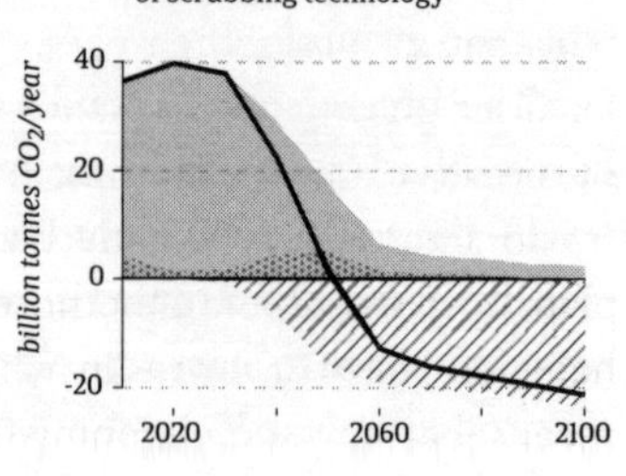

Source: IPCC, *Global Warming of 1.5°C: Summary for Policymakers (2018)*, p. 16

And now for the disastrous news. We don't even know yet whether it's possible to process a CO_2 surplus with the help of biofuels and underground storage.[35] There's no certainty, for instance, that once pumped underground the carbon will remain there.[36]

And the technology for doing all of this is extremely expensive at present. To achieve 'negative emissions' on the required scale we'd need at least 3,000 large carbon capture and storage factories by 2070.[37] The construction of these facilities would cost tens of billions, money that could be saved if we prevented the emissions in the first place, instead of cleaning them up afterwards.

The most problematic aspect of this half-measure is the production of the biofuels themselves – the willows, poplars and other crops intended to be burned in the power stations. In the standard plans developed by the IPCC, this would require converting an area once or twice the size of India into biofuel plantations.[38]

That's an inconceivable size. In the worst-case scenario we're talking about nearly *half* of today's global arable land, which is already being depleted by industrial agriculture.[39] This is land that isn't going 'spare'. Because fertile land is in limited supply, the development of new biofuel plantations will almost certainly result in further deforestation and the destruction of nature reserves.[40] The CO_2 released by logging could undo the 'gains' of using biofuels before we've even started, as was the case with the palm oil plantations.[41]

Besides, the production of the necessary biofuels could double the demand for chemical fertiliser – which would also lead to higher emissions, because the manufacturers of chemical fertiliser often burn natural gas in their plants.[42] In turn, excessive use of chemical fertiliser on the biofuel plantations could lead to higher emissions of nitrous oxide,[43] something that's already happening.

In short, this is a truly bizarre, ill-considered and dangerous plan, born of desperation: those who can't kick the oil, gas and coal habit dream of a future in which we can turn back the clock. It means that we're condemning future generations to the production of biofuels on a shrinking area of farmland.[44] And all because we're putting off dealing with the problem.[45]

*

The politicians in their economic straitjackets who arrive at these three half-measures are avoiding major political decisions in the short term, opting instead for cheap measures that won't offend anyone in their home country. Visionary interventions are for other people in other times. As a result, we all end up being worse off.

But even if this non-committal attitude prevails in the decades to come, we still won't find ourselves in the future I described at the start of this chapter. To get there, in a world in which global warming mainly affects the poor while the rich are still doing pretty well for themselves, some of the richest countries and individuals would have to deliberately wash their hands of the problem – in ways more obnoxious than apparent in the well-intentioned but ill-conceived half-measures I just described. Unfortunately, that's a development we're already starting to see.

The super-rich are drawing their conclusions

Some political leaders behave as if they're living on an altogether different planet. Donald Trump is a case in point. In June 2017 he announced that the US was withdrawing from the Paris climate agreement. As far as Trump was concerned, America would continue to blast its CO_2 into the atmosphere, and how that would affect the rest of the world wasn't his problem. He doesn't believe in climate change anyway.

But if Trump wanted to flout the climate rules he might as well have stayed *in* the Paris Agreement: the climate accord doesn't contain a single binding rule and no concrete measures for limiting emissions. It's made up of intentions. The idea was that countries would come up with their own plans to meet the targets, and that they would spur one another on to be more ambitious. Now that the world's leading power has shown that it's fine to ignore the climate, other countries are also watering down their plans. Since Trump's farewell to 'Paris', big oil-extracting countries such as Russia and

Saudi Arabia are taking the liberty of sabotaging the annual UN climate negotiations.[46]

It's tempting to pretend that President Trump is a special case, but he isn't by any means.[47] He fits neatly into a tradition of American Republicans subverting climate policy. His predecessor, George W. Bush, withdrew the US from the Kyoto agreement that was aimed at curbing emissions. Now Trump is even giving the powerful American fossil fuel companies what they want: every opportunity to extract new oil and gas.

In Trump's wake, we're seeing the rise of 'strong men' the world over, all keen to make their country 'great' again. None of them have faith in the power of international cooperation. To them, the world is a zero-sum game – my gain is your loss.[48] Trump said so literally in June 2017: 'This [Paris] agreement is less about the climate and more about other countries gaining a financial advantage over the United States.'[49]

Put differently: working together to protect the environment is not an option – if I impose restrictions on my economy, another will reap the rewards and my country will lose power.

*

In a world with leaders such as Trump at the helm, it's every man for himself. And so we're seeing a trend in which those with plenty of cash dream up ways to avoid the worst consequences of climate change. Some billionaires – Elon Musk, Richard Branson and Jeff Bezos – have expressed the wish to colonise space. But the same escape fantasy is reflected down here on earth as well, in the construction of bunkers and private enclaves.

Take the Survival Condo Project. In an abandoned underground rocket storage facility in Kansas, project developer Larry Hall has had fifteen luxury apartments built. The bunker has armed security at the door; only people with a pass are allowed to enter in the event of a crisis. The former missile silo has enough food and fuel for five years. Work is being done on a renewable energy supply, the cultivation of vegetables under LED lights and an indoor fish farm.

With 'half-floor condo suite packages' starting at $1.5 million, a select group of Americans has reserved a place here in case society crumbles after a sudden disaster or a catastrophic confluence of climate change and other crises. If necessary, an armoured SUV can come and collect the residents. Inside they can avail themselves of all mod cons: there's a swimming pool, a gym, a cinema, a dentist and even a shooting range. Large LED screens on the walls show panoramas of choice in lieu of windows: the prairie, a pine forest or Central Park in New York.

The Survival Condo Project is not unique. In a former bunker in Germany, for example, work is underway on what business magazine *Forbes* describes as an 'invitation only, five star, underground survival complex, similar to an underground cruise ship for the elite'.[50] Many buyers have a helicopter or a private jet at the ready to take them to their place of refuge.

But not every super-rich doomsday prepper opts for a bunker; remote islands are popular, too. Peter Thiel, the co-founder of PayPal and one of the world's richest men, has bought a 193-hectare estate in New Zealand he can fly to if and when civilisation collapses. He's one of the hundreds of Americans who have applied for a New Zealand passport, as the country is seen by rich preppers as the place to go since it's so remote and beautiful.

Gated communities likewise reflect this flight fantasy. The principle of entrenchment – a group of rich people erecting a residential neighbourhood, building a wall around it and hiring their own security – is obviously far from new. But in response to the looming climate crisis this type of enclave is gaining in popularity, as well as in scale.

In Nigeria, for example, building is underway on Eko Atlantic, a gated community on an artificial island off the coast of Lagos, the country's biggest city. Eko Atlantic is meant to become Africa's economic centre: a Mecca for rich Nigerians, jetsetters and multinational headquarters. Its two private developers are promoting it as a 'sustainable city, clean and energy efficient with minimal carbon emissions'. The complex is due to house 250,000 residents and provide employment to 150,000 people – and these will be the only individuals with access to the island. The sole aim of Eko Atlantic, the developers say, is 'to arrest the ocean's encroachment'.[51] The island is on higher ground than Lagos itself, and erected around it is a wall of 100,000 heavy concrete blocks to protect against sea-level rise.

In stark contrast, 70 per cent of the more than 20 million inhabitants of Lagos itself live in shanty towns. There's no concrete wall to protect them from further sea-level rise. Nor is there any luxury on the scale of Eko Atlantic: in much of Lagos, the power supply is unreliable and running water intermittent.

Walls are the common thread in this future scenario, and not just in poor countries. Plans have been drawn up for a massive wall around Manhattan that should protect Wall Street from the rising sea level. The one disadvantage: other boroughs could face more flooding as a direct result.[52]

Safety for the chosen few

It stands to reason that it's mostly the poor who are hardest hit by warming. Look at what happened in September 2017 when Hurricane Irma descended on Miami, the balmy city with the beautiful beaches at the southernmost tip of Florida.

That's when two Miamis emerged.

There was the Miami of the rich, who were well prepared. One of them was 25-year-old Matthew de La Fe. Just before the storm made landfall, he was interviewed while using some exercise equipment on a local beach. It was quiet there, because most people had been evacuated.

De La Fe was confident that he'd be fine: his father had built their house on a hill and they had their own generator in the event of power failure. 'I'm planning to give Irma the finger,' he told a reporter with the *Guardian*.[53]

Then there were the people who couldn't afford extra roofing or hurricane-resistant windows – they could barely even pay their rent. One of them was 55-year-old Michael McGoogan from Liberty City, a poor neighbourhood. He worked in a food-processing factory and was renting a room in a house with eight others.

When McGoogan and his fellow boarders spoke to their landlord about the storm, they were told to sort it out themselves. The group managed to gather enough timber to board up two of their ten windows. They could only hope for the best. 'If the windows start smashing, I'm going to put my mattress up against them and if that don't work, I'm going to hide in the bathroom,' McGoogan said. 'We got what we got.'

The hurricane derived its destructive force from the warmer ocean and caused tens of billions of dollars worth of damage in Florida.[54] While it didn't take long for emergency aid to get underway in Liberty City, much of it organised by locals themselves, the residents of this poor area soon woke up to a new ordeal: without their knowledge the city council had designated several vacant blocks in their neighbourhood as landfill sites. Overturned trees, mattresses, refrigerators and other waste from all over Miami was dumped there. The debris piled up, several metres high.[55]

Rats, snakes and raccoons were having a ball.

The people less so. In neighbouring Monroe County the number of suicides doubled in the first half of 2018. 'It's like this miasma, this cloud hangs over anybody that had to go through it,' the director of a local healthcare facility was quoted as saying.[56] Depression and post-traumatic stress are not uncommon after storms, floods and extreme heatwaves.[57]

More than a year later, many people in Florida were still battling insurance companies to get compensation for the damage. I've been unable to trace Michael McGoogan. But Matthew de La Fe was doing well, if Instagram is to be believed. He was studying for his Bachelor's in Business Administration and travelling around America. In August 2019 he did his first half Iron Man – a long-distance triathlon.

*

All this shows that global warming is already exacerbating the disparities between the haves and the have-nots. And history shows us that this inequality leads to tension and conflict.

In 2011, the civil war in Syria was triggered in part by years of drought that drove 1.5 million people from their land. They migrated to the cities, where a shortage of social amenities and a spike in unemployment and criminality caused growing tensions. Drought was obviously not the only issue that led to the conflict, but it was certainly a contributing factor.[58]

Migration and conflict as a result of drought, rising food prices and food shortages have been on the increase in recent years, especially in countries where the social fabric was already strained – think of Pakistan, Burundi and Somalia, as well as Nigeria and Iran.[59]

At present, a total of 70 million people worldwide have been forced to leave their homes because of violence, hunger or another catastrophe in their immediate surroundings. And the number of involuntary migrants is rising. A further thirty-one people are newly displaced every minute.[60] The vast majority of these displaced people remain in their own country or find shelter in the region. A small percentage embarks on a longer journey, which far too often ends in tragedy.

In recent years, hundreds of migrants trying to reach Europe have lost their lives in the Mediterranean. In response to the 'refugee crisis' that broke out in 2015, Europe has tightened its border controls. The EU has agreed a deal with Turkey and is training the Libyan coastguard to block migration routes. Thousands of migrants are being detained under inhumane conditions on the Greek islands of Lesbos and Samos.[61]

In Europe we like to think that we're much more civilised than the orange-tinted US president with his border wall with Mexico. But the European response to the problems elsewhere isn't fundamentally different from Trump's. While people beyond our borders are facing hardship, we're opting for extra security to keep them out.[62] It has come to the point where the Mediterranean is the deadliest place on earth for migrants.[63]

As the sea level rises and floods, droughts and severe storms increase, worldwide migration flows are expected to do the same. The International Organization for Migration puts the number of climate migrants at 200 million by 2050.[64] Drought and food shortages drive people to despair, experience tells us, and desperate people create unrest. In anticipation of this, defence forces around the world are preparing for an increase in conflict. The Pentagon, for example, views warming as an 'urgent and growing threat' and as a 'significant risk' to US security.

Potential flashpoints are already emerging: areas where food and water are scarce and becoming scarcer still. Ethiopia and Egypt are already embroiled in a dispute over the use of water from the Nile. In the unsettled border area of Pakistan, India and China, the drying up of rivers, which are currently still fed by glaciers from the Himalayas, will cause tensions to mount between the nuclear powers.[65]

What would it be like?

If we carry the developments I've just outlined through to the future, we can begin to see how we might actually end up in the world with which I opened this chapter – a world of 'red zones' and 'green zones', one in which security and comfort are extremely unequally divided between the winners and the losers of climate change.[66] What would it be like to live in this future?

Economic decline. The global economy will be under immense pressure. Heat, drought, more severe storms, floods and fires are already costing the world economy hundreds of billions per year,[67] and this figure is due to increase.[68] But it's not just the climate that sustains damage; economic growth itself suffers from warming.

In 2015, researchers for the science journal *Nature* looked at the link between the average temperature and economic performance in terms of gross domestic product (GDP) in 166 countries between 1960 and 2010.[69] What they found, one of them said, was 'almost like a law': economies perform best around an average of 13 degrees Celsius.[70] In recent centuries, countries with a moderate climate – the US, Japan and large parts of Europe – have been close to this optimum. Higher temperatures see a decline in agricultural yields and productivity levels, among other things, and that depresses growth.[71]

This means that further warming will create winners and losers. Countries that are already warmer than 'optimal' for economic development – among them India and Brazil – stand to lose even more in the event of continued warming. In South Asia, 800 million people would see a decline in their standard of living.[72] In the US, each additional degree would depress growth by 1.2 per cent.[73] In countries that are currently slightly colder than 'optimal' – such as Finland and Canada – the economy will actually improve a little.

But in real terms, everyone will lose out in this future, as 90 per cent (!) of the world population lives in countries where the economy is expected to perform worse as the climate continues to warm. This could make the world between 15 and 30 per cent poorer by 2100 than it would be in a future without further warming.[74]

The exact damage to global output can't be predicted – the figures are shrouded in too much uncertainty. But the studies do suggest where we're headed: higher costs and lower incomes. And while earning less, we'll have to spend more money on adaptive measures such as upgrading urban drainage networks to deal with heavier precipitation. These are costs that are chronically underestimated.[75]

No sustainability. It's all but certain that we'll still be using large quantities of fossil fuel in this future. The investors who are pumping $1 trillion a year into new pipelines, drilling rigs and oil tankers all want to see returns. And they'll only do so if this infrastructure is used as originally intended: for the extraction and transportation of oil and gas. So that's the trade-off in this future: fossil investors are still earning their money, while the rest of the economy is straining under the pressures of a hotter climate.

It's likely that renewables will continue their ascent in this future, but they still won't be able to compete with their better-financed rivals. Likewise, farmers who want to innovate sustainably are toiling in the margins. They're unable to keep up with their subsidised colleagues. At best, they'll exist side by side, the way biofuels and CO_2 storage exist alongside the fossil fuel economy.

If the historical pattern prevails – which sees those in power opting for cheap measures that won't rub anyone in their own country up the wrong way – then the production forests for biofuels will spring up in relatively poor countries: Brazil, Indonesia, Nigeria and the Democratic Republic of Congo are already showing an interest in the development of such plantations.[76] Richer nations will import the wood pellets, pump the CO_2 into the ground and become, at least on paper, 'climate neutral'. In the supplying countries, meanwhile, the food and water supply will come under increasing pressure, while ever more forests disappear.

Colonial relationships, back with a vengeance.

Strained relationships. What exactly you can look forward to depends on where you were born. If you live in a relatively rich enclave – whether it's a gated community, London or Fort Europe – you may still be doing okay. Your freedom to do as you like will be sharply curtailed: holidays in less privileged parts of the world are no longer on the cards, and national service may be reintroduced to keep the borders safe.[77] But all things considered, you'll be doing reasonably well.

If you were born outside a rich enclave, you've drawn the short straw, even more so than today, since the number of conflicts revolving around fresh water, food and raw materials will have risen in this future.[78] If such a conflict breaks out somewhere, there may be an overriding urge to quell it with military force – anything not to lose control. Authoritarian leaders are likely to thrive in this world. They will blame every new crisis on 'the other' and fantasise about a homogeneous 'we' or 'us' who will 'beat it' with force, unity and closed borders. You could see precisely these impulses on full display during the corona-outbreak. Donald Trump was the bluntest example as usual: he called Covid-19 a 'foreign' and 'Chinese' virus, casting it as a foreign invader, then opted for isolation. Every man for himself. It didn't work.

Solidarity with people who are less fortunate may be in short supply. As it is today. When Hurricane Irma had reduced all of Sint Maarten to rubble, one publicist wrote on Twitter: 'You don't *have* to live on an island where every few years a hurricane destroys everything.' As if the average income in Sint Maarten is high enough for its residents to up sticks and leave.

So what this scenario shows: if we continue down our current path, we're sure to end up with higher temperatures, higher sea levels and greater ecological decline than we've ever had to contend with.

Don't forget, the climate is merciless. If the winters are no longer cold enough for the creation of (lots of) new ice on Greenland, the warm summers will get the upper hand and the fate of the ice cap will be sealed.[79] In the words of journalist Bill McKibben: 'When your ice caps are melting fast, winning very slowly is another word for losing.'

'To be honest, I've kind of given up on the earth,' one of my friends says when our conversation turns to the climate. 'We'll never be able to put a halt to climate change. People want to, but our juggernaut of a system charges ahead.' In this first scenario I've just outlined, he's right: we're slowly slipping into the zero-sum world of the Trumps on this earth. Whether we like it or not, we're sliding towards a hothouse earth – that's the story we're going to have to tell our children in this future.

Is there an alternative? There certainly is. Welcome to the 2050 of your second possible future.

Chapter Six

FUTURE SCENARIO 2: FORESTS

Year: 2050
Somewhere in Western Europe
Global warming since the Industrial Revolution: 1.6 degrees Celsius

You can see the consequences of the Great Turn everywhere. The city is no longer a concrete jungle. Green roofs have been constructed on houses to absorb CO_2 and rainwater and cool the surroundings; parks are home to a wider variety of plants and animals than ever before.

The turn began in the 2020s, when more and more governments levied taxes on CO_2 emissions and invested the profits in renewable technology. The green industries that flourished under that policy created millions of new jobs and substantially boosted the economy. We're now producing green energy, eating sustainably cultivated food and almost all everyday items are produced in a closed loop.

The turn came at a hefty ecological price: countries such as Bolivia and the Democratic Republic of Congo have expanded their polluting mining industry in recent decades to produce sufficient resources for wind turbines, batteries and modern power networks. But now that we manage the recycling of our materials better, mine closures are on the increase. The definitive end of the long era of plundering is finally in sight.

Global warming has stopped, although in the past decades it has grown warmer – part of the damage was unavoidable, thanks to the CO_2 from previous generations still floating around. Despite the turnaround, in the past decades there have been thousands of victims worldwide as a result of heat, drought and other climate disturbances.

But the climate fund to which all countries contribute a fair proportion of their GDP has mitigated the pain in the worst affected areas, and the collaborative efforts of the national Recovery ministries are bearing fruit: communal forestry projects deliver tangible benefits for everyone. A great deal of money that previously went into defence now goes to the restoration and protection of nature. The millions of new trees planted in the 2020s now form forests that retain carbon and provide cooling during the warm summers. In spite of everything, it is a hopeful time to live in.

Faith in the future has returned.

*

This second future scenario is a real alternative to that of the previous chapter. In this future world, global warming has been effectively mitigated and the wealthiest countries choose to help the poorer countries. Governments place the emphasis on recovery and innovation in order to live with the consequences of the warming that could no longer be avoided. This, by contrast with the first scenario, is a world of collaboration.

In order to end up in this future, a great deal needs to change in the next thirty years. Not in one go, but step by step, at a rapid pace. Such a turn will bring substantial costs and downsides. I'll return to these at the end of this chapter.

First I'll describe what's needed to set the turn in motion. As in the previous chapter, this is not mere speculation: I'll show, based on current and historical developments, that the necessary ingredients for this future are also already present.

How technology can enable transformation

Let's start with technology. Currently, the entire economy is founded on fossil fuels. We use enormous quantities of them to generate electricity – for fridges and washing machines, phones and computers. We also use large quantities of fossil energy in applications from boilers to aeroplanes, from steel factories to petrochemistry. We use it to heat our houses and manufacture everyday items, to move us around and transport goods across the world. Chemical fertilisers, plastics and synthetic fibres – our fossil dependency is enormous.

Until recently, we had no credible alternative for this. But now that the necessary technology to replace fossil energy has been invented and developed, nothing stands in the way of us deploying this technology on a large scale. The complete replacement of coal and gas in the electricity supply is becoming increasingly feasible, as solar panels and wind turbines have started on a relentless growth spurt. They currently deliver only 3 per cent of all energy worldwide, but their use is growing faster than even the most experienced analysts thought possible. The prominent International Energy Agency, for example, predicted twelve times that the growth of solar energy would stagnate, but it still hasn't happened.

That's because solar panels and wind turbines are not *fuels* but *technologies*. It might sound blindingly obvious, but the difference is crucial.

For an economy that relies on fossil energy, you need to keep on extracting coal, oil and gas – ad infinitum. Because the reserves that are easiest to extract have been exhausted in many places, more money, energy and time, relatively speaking, are required to extract new fuel.[1] Once it has been used, waste is left behind (CO_2, soot, ash).

Solar panels and wind turbines are radically different.[2] They are produced and installed once. After that they supply energy for twenty to thirty years. Maintenance costs are low, the fuel costs nothing, and it's also infinite – as long as the sun keeps shining and the wind keeps blowing.

Add to that the fact that the manufacturing costs of the panels and mills keep on dropping. For every doubling of the total installed generative capacity, the price drops by around 20 per cent.[3] Thanks to the constant reduction in price, living off sun and wind becomes ever cheaper. It takes on average just a year to build a large wind farm, and just three to six months to install a solar farm. That means the proportion of renewable energy can grow fast, resulting in a rapid drop in emissions.[4]

The wind energy available per day is a good ten times the current

The solar panel revolution

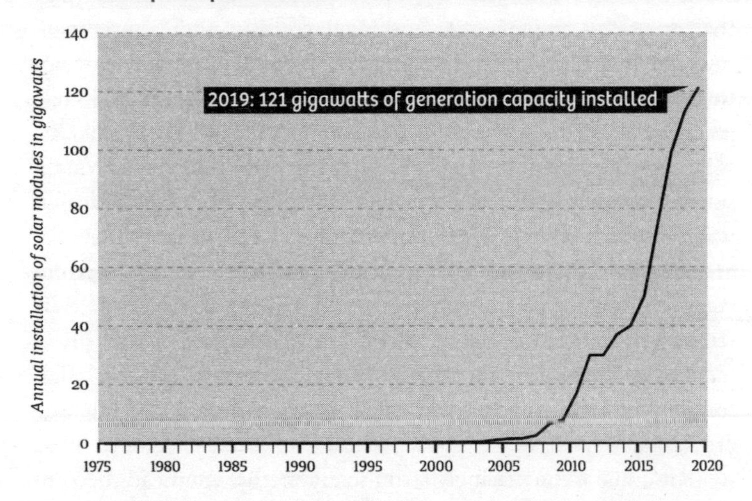

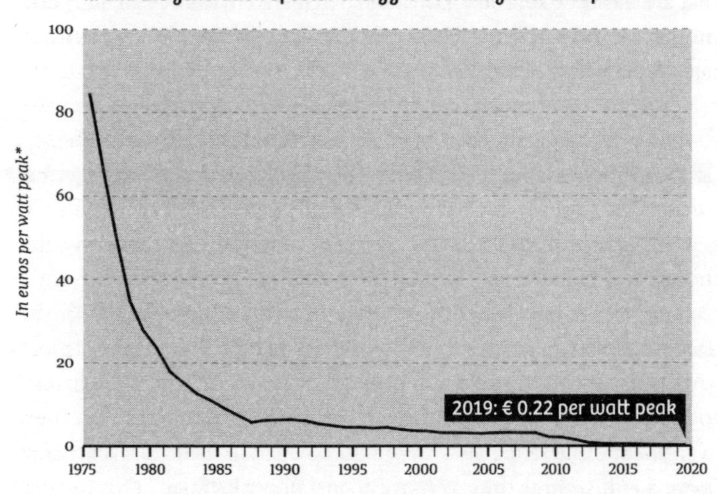

*One watt peak is the production of 1 watt (W) under standard conditions. Price per sun cell.
Source: Auke Hoekstra & Naud Loomans based on ITRPV & BNEF (the 2019 numbers are BNEF's best estimates on 20 January 2020).

total daily energy usage worldwide. And in an hour the world receives enough solar radiation to supply humankind with energy for a year.[5]

If we succeed in collecting that energy, we've struck gold. That's why so many companies are working to come up with new technology to harvest it. They're investing, for instance, in the development of kites that float in the air, attached to a cable, and produce an electric current by rolling a pulley back and forth, or in concentrated solar power, using thousands of mirrors to gather the sun's rays and focus them on a single point in a high tower. The intense heat created can then be used to produce electricity or stored for later use.[6]

Besides solar panels and wind turbines, there are other promising options for clean energy generation, such as wave and tidal energy (from the motion of water) and geothermal energy (from heat deep underground). You occasionally also hear calls for building new nuclear power stations; it's often forgotten that building a nuclear power station takes at least two years, but more often ten and, due to increasingly strict safety requirements, it's becoming more expensive in many places.[7] That makes it an impractical, expensive, slow solution – the construction of the Flamanville plant in France and a new reactor at Hinkley Point in the UK are painfully apt illustrations of this.[8]

Not that that means we should close existing nuclear power stations early: after all, they provide much-needed CO_2-free energy. If innovation makes it possible to build safe, new nuclear reactors, those who care about the climate shouldn't dogmatically oppose it.

But the probability that a new generation of nuclear power stations can compete with solar panels and wind turbines is currently rather low: it has very little chance now of catching up with the prices of energy from sun and wind. In 2017, the total investment in new renewable generation capacity was already twice the investment in new coal, gas and nuclear power stations put together: a sign that investors, too, have come to the conclusion that solar panels and wind turbines offer the best opportunities.[9]

Until recently, the variability of sun and wind was an obstacle to upscaling: the great fear was that we would have no electricity on still or cloudy days. But such fluctuations in supply and demand have proven technically easy enough to balance out by installing more powerlines, so that regions with a surplus can service regions with shortages – it's never still or cloudy everywhere at once. For households and neighbourhoods the daily fluctuations in supply and demand can be cushioned with lithium-ion batteries, which have become 80 per cent cheaper in the past eight years,[10] and are expected to continue to fall in cost.[11] With the proven reserves of lithium, we can make do for another 200 years,[12] and work is in progress on a new generation of clean batteries made from sand – we have enough of that in any case.

In winter, when the sun doesn't shine much and we need to use lots of energy to heat our houses, we can make use of 'solar fuels': clean fuels that energy companies plan to produce on a large scale at moments when there is a surplus of green electricity available. On an old oil rig in the North Sea, for instance, they're building a test factory for producing hydrogen. The process is simple: you pass electricity generated from wind through a container of water (H_2O), separating the H-molecules from the O-molecules and there you have it, you've made hydrogen. In winter, energy suppliers can stoke this clean fuel to generate electricity for all the houses and factories that need a little extra – just as we now generate electricity by burning natural gas.

*

You don't need to be a naïve tech-optimist to derive hope from all the renewable alternatives we have available or in the pipeline (a pipeline containing hydrogen, that is). With a combination of powerline connections, batteries and solar fuels, we'll still be able to guarantee a continuous energy supply in the future.

Did the corona-crisis change any of this? Not really. Of course, lockdowns caused delays and disruptions along the supply chains of solar panels and wind turbine manufacturers. They also made it very hard to plan or construct new large-scale renewable energy projects, which will probably curtail the growth of renewables in 2020. But previously planned projects were finished, the sun shone and the wind blew, and so even during the corona-crisis, the production of renewable energy increased – as demand for oil, gas and coal plummeted. Soon, people started talking about a 'green recovery' and speculating that the switch to renewables could actually pick up speed in the aftermath of Covid-19.[13]

But obviously, the so-called 'energy transition' isn't a done deal yet. The problem is, we use fossil energy for much more than electricity generation. But here, too, new technology beckons: we can heat houses with electric heat pumps instead of central heating boilers, cars can be driven by batteries instead of combustion engines, steel can be recycled with electric-arc furnaces or newly made with hydrogen instead of coke (a coal product).

The common denominator in all these technologies is that they consume electricity instead of oil, gas or coal. If that electricity is renewably generated with solar panels and wind turbines, that delivers gigantic climate gains.

These technologies are not yet perfect: the purchasing costs of heat pumps, for example, are still high, the quality of the devices varies considerably from one provider to another, and recharging an electric car isn't yet as simple as filling up a petrol car. But people are working hard on improvements.

The plastics we currently still make from oil could be manufactured using algae in the future.[14] Eliminating oil from air and sea transport will be more difficult again; the first technologies exist, but they're still in their infancy.

The most important point is that we already have the technologies to make the energy system radically more sustainable. We're just not using them nearly often enough.[15] We need to bring them in at least six times as fast to reduce CO_2 emissions and achieve the 2-degree goal.[16] At the current pace, the energy supply would only be fully cleaned up in 350 years.[17] Optimism about technological innovation isn't enough.

But it is fantastic that the clean alternatives exist: it means that we can erase the biggest source of our emissions, without necessitating a decline in energy consumption.[18] If we *do* save energy, then the challenge we face becomes a great deal more manageable.[19] But 'less' is no longer the only choice: 'better' is now also a credible option. That was still unthinkable ten years ago.

In agriculture, the second-biggest contributor to global warming, great innovations are also taking place. People have developed (or rediscovered) many methods to produce sufficient food for the growing world population without using pesticides or artificial fertiliser. One of the most promising perspectives is offered by agroecology.[20] The farmers at the forefront of this farming method look for ways to cultivate multiple kinds of food simultaneously – with different plants and sometimes also animals complementing one another.

Agroforestry is a good example of this approach. Instead of extensive fields with long rows of identical crops (monocultures), farmers using this method attempt to imitate forests, with a source of food on every 'storey'. Root veg in the soil, for example, and apple trees overhead. Or cows on the ground level and nut trees above.

Because the trees and shrubs stay put in a multifunctional field of this kind – only the yield is taken from the land – the soil life becomes ever richer and local biodiversity increases.[21] Birds, soil organisms and insects, along with fungi and bacteria, form a natural defence system: lice and caterpillars, treated with pesticides in industrial agriculture, are ideally eaten by birds here.

Given that a farmer can produce multiple types of food on one stretch of land, agroecological farming uses energy and space far more efficiently than industrial agriculture: you don't have to put as much in and you get more out.[22] Sowing, fertilising and spraying don't have to happen every year. And with the trees staying where they are, the carbon they filter out of the air remains safely in the wood and in the soil. Agroforestry can thus lower the CO_2 concentration in the air – and hence help moderate global warming itself.[23] (The general name for this practice is 'carbon farming'.)

Of course, in order to arrive at a sustainable world, we certainly need more than a transformation in the areas of energy and agriculture. We need to make the switch to a circular economy, in which more and more material and energy cycles are closed, and waste becomes a thing of the past. That too is possible.

In the clothing industry, fabrics such as hemp, flax and nettle can provide sustainable alternatives to cotton, polyester, wool and silk. Transitioning to these sustainable materials reduces water consumption, emissions of greenhouse gases and use of fertiliser and pesticides by this sector.[24]

In construction, we could make much more use of bamboo, an extremely fast-growing grass species that fixes CO_2, and that has the compressive strength of concrete and the tensile strength of steel.[25] Hemp turns out also to be an excellent building material. And of course there's wood: one of the most sustainable ways of keeping carbon out of the air is to store it in wooden buildings.[26]

Just under 5 billion tonnes of cement produced annually worldwide are the cause of a good 5 per cent of all CO_2 emissions. By changing the composition of cement and more often opting for alternative materials, these emissions, too, could drop rapidly.[27]

All these innovations together offer the prospect of a future in which we break out of the vicious circle of continually extracting, emitting and exhausting resources. It will require gigantic changes which can't be precisely outlined in advance. But we don't need to know all the answers before we begin: people have always learnt by doing, by experimenting, making mistakes and trying again.

The most pressing question now is: how do we ensure the desperately needed increases in scale and pace?

The rediscovery of the entrepreneurial state

For a mass upscaling of clean technology, the world economy needs to extricate itself from fossil energy quickly. That will only be possible if governments play a very different role from that in the first future scenario, and throw their full weight behind the necessary transformation. Is that possible?

History suggests it is. An example is the action of the American government during the Second World War. In 1941, the biggest factory of its time was built in six months in the state of Michigan. Within two months, it was manufacturing a B-24 Liberator bomber every hour. A complicated aeroplane with more than 1 million parts in an hour – 'endlessly more intricate than solar panels or turbine blades', writes Bill McKibben.[28]

This happened throughout the country. The Pentagon had companies compete to step up production of military equipment as fast as possible. Companies that didn't want to be commissioned by the government were compelled to participate, or otherwise close down.

All kinds of unexpected combinations suddenly turned out to be possible, according to McKibben. 'The company that used to supply fabrics for Ford's seat cushions went into parachute production. Nothing went to waste – when car companies stopped making cars for the duration of the fighting, GM found it had thousands of 1939 model-year ashtrays piled up in inventory. So it shipped them out to Seattle, where Boeing put them in long-range bombers headed for the Pacific.'

My comparison with American wartime manufacturing is of course extreme: such borderline authoritarian state action is not what we prefer in a modern democracy. But the fact that such collective mobilisation has been achieved in the past does show that a slightly less extreme variant is at least *possible*.

The coronavirus showed this to be true. Suddenly governments demonstrated their ability to listen to scientists, free up cash and take the lead. Even while the scientific facts on Covid-19 were much less robust than those on climate change, politicians decided to act. The responsible ones among them acted with precisely the 'better safe than sorry' mentality that we need to stop the habitability of the earth from shrinking further. In the words of the German chancellor Angela Merkel: 'It is thin ice, a fragile situation, a situation where caution is the order of the day, not overconfidence.' Right.

Budgetary rules were let go of, central banks rushed in to help and money was made available to the researchers who needed it most. The speed with which testing capacity and intensive care capacity was ramped up reminded many of wartime. But it was peacetime, and we were learning something about our capacity for collective action.

*

Another recent example of persistent government action is the development of solar energy. In 1954, it was Bell Labs, then the research laboratory of telecommunications company AT&T, that invented the first modern solar cell. But it took the US Department of Defense and space agency NASA to make the technology grow. The Americans were enveloped in a space race with the Russians and needed nothing more urgently than solar cells for their space satellites. They threw money at the technology until it had matured enough. In 1958, the first satellite to use solar energy was launched from Cape Canaveral, Florida. The foundation for the modern solar industry had been laid, on government cost.

Or take the towering wind turbines, which along with solar panels are at the forefront of the current energy revolution. They too are the result of government investments. The oil crisis in the 1970s spurred on the US, Germany and Denmark to make substantial innovations.

What is remarkable is that these countries didn't come to any explicit mutual agreement, but still, between them, encouraged the development of wind turbines. Demand in the American state of California created security of investment for the Danish company Vestas, currently the world's largest manufacturer of wind turbines. Vestas was able to grow big because it had bought up the patents for turbines developed by researchers in government-funded Danish laboratories early on.[29]

Historically, it has often been governments that made the breakthrough,[30] or even individual politicians. One example is Social Democrat Hermann Scheer: a visionary advocate of renewables in the German parliament. In the year 2000, he ensured that a law was adopted ruling that owners of wind turbines and solar panels in Germany received a guaranteed purchase price for the electricity they generated for fifteen years.

Suddenly it was profitable for individual citizens to go into energy generation. The demand for panels and wind turbines exploded, as did the generation of green electricity. In 2000, 6 per cent of German power was renewable; now the figure is over 40 per cent.

German politics intentionally created a new market to spark innovation. 'We are responsible for this problem [of climate change] and should not shrug this responsibility by only thinking of ourselves,' said Scheer in 2000.[31] The subsidised development of renewable technology was Germany's 'gift to the world', an energy analyst later commented.[32]

The German government was willing to invest in the future, and the consequences of this policy reached far beyond its borders. China soon saw that the German policy offered a big opportunity; it meant they could be certain that newly manufactured solar panels would be purchased. The Chinese government decided to invest in talent development, research and mass production. China is now the world's biggest manufacturer of solar panels, and the panels are a fourteenth the price of those in 2000.

Emissions from the German energy supply dropped: the deployment of renewable energy between 2000 and 2018 prevented 2.1 billion tonnes of CO_2 emissions in total – approximately equivalent to two years' worth of emissions from Japan, the world's third-largest economy.[33]

Scheer could never have foreseen how fast solar panels would develop. He couldn't be certain of the success of the subsidies he introduced. He did what he believed in – and the consequences were astronomical, because many others did the same.

The history of wind turbines and solar panels shows what economist Mariana Mazzucato has been saying since 2013: the state is the greatest inventor of all time.[34]

How states can compel breakthroughs

Far from all countries are as proactive as China, Denmark or Germany. But it's important to emphasise that the entrepreneurial state *has never disappeared*. Not even after the fall of the Berlin Wall, when everyone was singing the praises of the free market.

With all the subsidies, tax breaks and shares in fossil-energy companies, national governments are enabling the continuation of global warming. It's the licences the governments hand out that allow oil and gas giants to keep on drilling new reserves. Since fossil-energy companies with shareholders can rarely escape the grip of the current system under their own steam, the government should do what it takes to compel them. What do we need to make that happen?

Governments should at least *want* to do it. And they claim that they do. The twenty richest countries, the G20, pledged in 2009 to wind down subsidies for fossil energy to stop climate change. Not all countries have made good on that ambition, but important first steps have been made.[35]

Indonesia decided in 2015 to scale back subsidies for the use of petrol and diesel for consumers. Canada and Argentina have removed some tax advantages for fossil energy exploration and extraction in recent years. Germany, the Czech Republic and Spain subsidise coal mining, but for a couple of years now more than three quarters of that subsidy has been specifically intended to help miners find other work and close old mines.

For example, Spain planned to close twenty-six coal mines in exchange for an investment of €2.1 billion in the areas affected. Strange, you might reason, as in fact the Spanish government is losing the tax revenue for the mining of resources *and* has to spend money to mitigate the pain in mining regions. Why does Spain want to do this? Because the politicians elected in the country can see that renewable-energy sources already supply more than two thirds of all Spanish electricity.

It's an inexorable economic logic: renewable electricity is becoming ever cheaper, so there's no future for coal power stations. Closure is inevitable. Politicians can mitigate the pain for local mining communities.

Or to put it another way: because the production of renewable energy is becoming ever cheaper, a serious climate policy for governments is becoming ever more attractive. And that very government involvement in fossil fuels is what gives them influence. Every subsidy for oil, gas or coal can be turned into an investment in renewables. To counter a tax break, a CO_2 levy can be imposed. The World Bank's latest overview shows that in 2018, fifty-seven countries had introduced a form of CO_2 pricing, and that nearly 100 others were considering such a measure.[36] The price tag that states slap on CO_2 is often still much too low to compel real sustainability, but the basis for a fair CO_2 price has been established.

Is it conceivable that in the coming decades *more* governments will do *more*? Certainly. It may seem cruel to think of the corona-crisis as an opportunity but, frankly, it is. The shakeup to the world's energy systems provides a huge opening to fast-track much needed interventions – it gives states leverage. With oil prices low, for instance, it's much less painful to ramp up carbon prices. And in return for any bailout, a renewable strategy can be ordered. Returning to business as usual is not at all inevitable, nor is it what the majority of people want: in a survey conducted across fourteen countries globally in April 2020, 65 per cent of respondents agreed that 'in the economic recovery after Covid-19, it's important that government actions prioritise climate change'.[37]

And there are many other reasons to assume more will be done.

China, emissions world champion, is now the global leader in renewable energy generation.[38] Beijing's human rights record being what it is, there's no reason to be all cheery about China. But the country *is* rapidly cancelling plans to build new coal power stations,[39] for the simple reason that electricity generation with solar panels and wind turbines in many places is cheaper *and* causes less local air pollution. China, its investments along the Belt and Road Initiative notwithstanding, wants to have greenhouse gas emissions falling from 2030.[40] This they're doing right.

India, the world's fourth-biggest greenhouse gas polluter, also intends to become a global leader in renewable energy. In 2018, solar panels covered more than half the new electricity-generation capacity in the country – new coal power stations provided 'only' a quarter of that new capacity.[41] Solar power in India is already half the price of electricity from coal.[42]

Worldwide, new coal power stations are still being built, but the number planned for the future is now shrinking by the year,[43] amidst complaints that financing new coal power plants is increasingly 'challenging'.[44] The corona-crisis is now hastening the demise of the world's dirtiest fuel: in the first quarter of 2020, global coal demand fell by almost 8 per cent. But even in 2019, before lockdowns battered the world economy, researchers recorded the largest ever global fall in coal-fired power production.[45] In fact, that decline in coal consumption combined with the growth of renewables to 'flatten the curve' of energy-related CO_2 emissions in 2019.[46] You could say emissions were ahead of the curve.

The majority of new investments in renewable energy come from developing economies, not from rich countries.[47] China and India are now in the lead. They see it as a strategic advantage to become more independent from fossil fuel suppliers by establishing their own renewable energy generation, while many European countries are hesitant, and the US is intentionally digging itself in further with more and more boreholes.

But even the American climate policy, or lack thereof, is being undermined. Within a week of Trump announcing that he intended to withdraw the US from the Paris Agreement, thousands of American cities, states and companies announced that they would still deliver on their pledge to reduce their own emissions in line with what is necessary to keep global warming below 2 degrees.

You might dismiss that as an empty gesture – after all, expressing an intention is different from making good on it. But efforts on the part of California, for example, show that the progressive forces in the US mean business: the state not only wants all energy in California to be clean in 2045; the governor has also signed an order that *the entire Californian economy* must be climate neutral by then. If that ambition is realised, this means that the fifth-biggest economy in the world will have reached zero net greenhouse gas emissions before the middle of the century. That's massive.

And it's highly probable that the demand for solutions from cities and states will spur on further innovation elsewhere,[48] just as in the 1970s, when wind turbines were still in their infancy and an entirely renewable world energy supply seemed an impossible utopia.

Governments are now starting to set hard targets. Sweden, France, the UK, Scotland and New Zealand have enshrined in law that by 2050 they will cause no more emissions at all. Countries often don't yet know *how* they will realise these kinds of ambitions, but it has already been determined in law *that* it will happen. In Costa Rica, Ireland, Belize and New Zealand, drilling for new fossil fuel reserves is now prohibited. In France, all oil and gas production has to stop by 2040. Germany will phase out coal by 2038. The UK will ban the sale of petrol, diesel and hybrid cars by 2035.

It's not just states: many companies either want to become more sustainable or fully dedicate themselves to the issue. Countless start-ups are searching for and finding opportunities in the flourishing sustainable economy. Existing organisations which are not up to their ears in oil have every freedom to formulate ambitions and carry them out. More than 170 big companies – including IKEA, AkzoNobel and Apple – have already decided to transition to 100 per cent green electricity.[49]

Philips, for instance, has reinvented itself as a green company that will soon exclusively consume green energy, that takes back outdated medical appliances from clients and reconditions them, and in Europe it argues for the strictest possible environmental legislation. The company has received compliments from Greenpeace and even acknowledges that it has discovered countless advantages from the green path. 'In the end our strategy provides better access to the capital market,' an investor relations officer said.[50]

Under the right circumstances, even oil and gas companies can transform, as the story of DONG shows. This Danish company in fact derived its name from the fossil fuels that built its fortune: Dansk Olie og Naturgas. The company is now known by the name Ørsted. It's the biggest installer of wind turbines in the world, and has divested itself of the majority of its fossil fuel activities – Ørsted's last coal power station will close in 2023 at the latest.

Why has Ørsted made this transformation? Chairman Anders Eldrup was convinced that his company must change completely within one generation in order to spare the climate. With a couple of colleagues who, like him, believed in that necessity, in 2008 he developed a green strategy. The group had to convince the many sceptics within the company *and* the government, which held more than half the shares in the company. Politicians were initially averse to the risks of the green path, but the company continued to focus on the new business.

It paid off. In 2018, Ørsted achieved a return of more than 30 per cent on the invested capital.[51] By comparison: the return on capital invested by Shell in 2018 was 9.4 per cent.[52] In recent years, companies leading on renewables on average have been doing better on the market than companies from the fossil fuel sector.[53]

*

Will all these developments be enough? As I wrote in the previous chapter, gradual reduction of emissions will not suffice. In order to keep global warming well under 2 degrees and effectively shrink the risk of passing tipping points,[54] CO_2 emissions would have to be halved every decade from 2020, so that we would be down to almost zero in 2050. However great the rise of green energy, however strong the promise of innovation in agriculture and however hopeful the ambitions of many countries, it's still not enough to achieve such a rapid drop in emissions.

So what can we do? We need to see an acceleration in the pace with which countries realise their green plans. History suggests that this really could take place. Every time people like Hermann Scheer make use of their limited room for manoeuvre to precipitate change, that leads to further change. The developments which people set in motion are thus the mirror image of the chain reactions that can emerge because one consequence of global warming leads to another.

Now that over 40 per cent of German electricity is renewable, the country is actively working on a completely clean energy supply by 2050.[55] China is confident it can end the smog which is making citizens sick and dissatisfied. Oil company Shell once summed up the law behind such developments in one of the future scenarios the company published: 'Reform unleashes new economic productivity and increases aspirations for further reform.' So Shell – obviously not a direct stakeholder in an accelerated end to the fossil era – also foresees that future change can go faster than one might expect based on results from the past. How we can achieve a 'great acceleration' is the subject of the following chapters. Let's assume here that that acceleration does take place. How would the world then look in 2050? Where do we end up?

How might things work out?

Economic costs and opportunities. The economy will probably do *much* better. Of course, revolution is expensive at first. Transitioning to agroecological farming, for example, requires big investments. But that's not the only cost of a turnaround.

If governments throw their full weight behind the development of a renewable energy supply, some countries and fossil fuel companies will be left with infrastructure they've invested in but which they can no longer earn as much from, if anything at all.

This could represent a serious threat to the world economy. According to a study in *Nature*, the total value of investments that could be left 'stranded' lies around the $9 billion mark. By comparison, the sudden depreciation in the housing market unleashed by the financial crisis in 2008 came to a 'mere' $0.25 billion.[56]

The big question is whether investors will withdraw their money from the oil and gas sector in time. The way things can fizzle out is illustrated by an example from recent history: the electricity market.

It's probably the biggest loss you've never heard of: €130 billion written off by electricity companies in Europe on their assets between 2010 and 2016.

Why? Electricity companies had completely underestimated the competition from renewably generated electricity from solar panels and wind turbines. Sun and wind were still relatively small competitors, but the energy they supplied was enough to capitalise on growth that the established electricity companies had been counting on. Their investments in coal and gas power stations swiftly became worthless and their market value began to nosedive.

While disastrous for those companies, it didn't result in an economic crisis. A significant proportion of investors *had* seen the risk coming and limited their losses by disposing of their shares. Other shareholders lost their investment, but the damage outside the sector itself remained limited because the 'correction' (i.e. drop in value) in the financial markets had already begun when the companies themselves denied that renewable power would become a threat.[57]

Things could go the same way if oil and gas companies encounter increasing competition from renewable energy sources: the markets may adjust of their own accord. Or not. In this scenario we would have to accept the risk of a financial shock due to sudden depreciations in oil and gas infrastructure, in order to prevent a greater risk – a hothouse earth.

What is certain is that the fossil fuel-exporting countries such as Russia and those in the Middle East and North Africa will be hit hard by the accelerated rise of renewable energy. But in the long term it offers these nations the opportunity to do what they've always said they wanted to do: become less dependent on a single export product or a small group of such products.

In this scenario, jobs would be lost in old industries – the 9 million jobs in the global coal industry, for example. That'll hurt. But a speedy transformation of the economy would also provide new jobs, potentially as many as 65 million by 2030.[58] Already, the green economy in the US employs ten times more people than the fossil fuel industry.[59]

Whichever study you look at, almost all economists agree that the benefits of the transition to a clean economy far outweigh the costs. That goes for the vast majority of individual countries, and for the world as a whole.

That's because there are so many benefits to a green transformation. From reduced spending on healthcare (cleaner air) to cheaper energy (the 'fuel' for solar panels and wind turbines is free and abundant).

Then there's the averted damage to the climate. With every tonne of CO_2 we *don't* emit, we save the world economy of the future an estimated $400.[60] The total benefit of stopping all emissions runs into the tens of billions per year.

That's the exchange in this scenario: fossil sectors lose, but the world economy wins.

Reduced inequality. In a future with cheap green energy, the differences between rich and poor will be much smaller. Twenty per cent of the world population currently has no access to a reliable power supply, let alone clean energy. Many people, especially in Africa and South Asia, have to cook their food on charcoal. More than a billion people still have no electricity. In the dark they have no light or use kerosene lamps.

Solar energy, combined with a lithium-ion battery and a small local electricity network, is a perfect alternative. That's now clear from various successful examples in countries such as India, Nepal and Peru. Inhabitants don't have to wait until their governments install a central electricity network or until a (foreign) company sees sufficient potential for investments.

Sure, there are start-up costs. You still have to purchase that renewable technology, after all. But microfinance and government investments might make that possible at ever lower costs. According to the International Renewable Energy Agency (IRENA), 60 per cent of the energy demand of the very poorest could be absorbed by the silent revolution of clean, local energy networks.[61]

And that's not the only inequality that will disappear. If we achieve a green transformation not only in the energy supply but also in food, consumption, construction and materials, the disproportionately large negative 'footprint' of the richest consumers will disappear. Up until now, proportionately, it's the richest people who are the greatest burden on nature. According to calculations by the French economists Lucas Chancel and Thomas Piketty, the richest 10 per cent of the world's consumers in the year 2015 caused 45 per cent of all CO_2 emissions. The richest 1 per cent caused as many emissions as the poorest 50 per cent.[62]

Crazy ratios, which disappear if we succeed in transforming the world economy. The old rule of thumb – that anyone who earns more also uses up more of the earth's resources – dissolves in this future, and the growth of the global middle class need no longer cause anyone environmental concerns.

And for those who worry about overpopulation, the average worldwide birth rate has already halved since 1950, and it continues to fall wherever women go to school and gain access to contraception. The UN predicts that at the end of the century there will be 10 to 12 billion humans, and that population growth will stop there.[63] There is no doubt that it is technologically and ecologically possible to sustainably feed and supply energy to a world population of that size.[64] In this future, the feelings of guilt some people now experience over the impact of having children on the climate will fade.[65] If we succeed in reducing emissions to zero by 2050, babies born now will be climate neutral most of their lives.

The rise of sustainability also has radical effects on geopolitics: in this future scenario there will probably be fewer conflicts than in the first scenario. In history, many wars have been waged over access to scarce resources, and the differences between poor and rich which have arisen in the fossil era are gigantic: the uneven distribution of oil and gas reserves in different countries has shuffled the cards for centuries of geopolitics based on abundance for some and scarcity for others.

We can now bid that geopolitics farewell.[66] On the one hand because the pressure on the planet in this future is considerably reduced, and on the other hand because now *everyone* will have access to endless supplies of clean energy. Roughly speaking, the northern countries with relatively few hours of sun will be able to make plenty of use of wind energy, while the less windy tropics have ample sun. Regional electricity networks will be linked up – a good basis for collaboration for the common good.[67]

The emphasis, therefore, in contrast with the first scenario, is not on what we have to defend. Border walls will be less necessary in this future. In this world there is more freedom than in the previous scenario, and that freedom is more fairly distributed.

Collaboration with nature. Our relationship with nature will change radically if we switch to agroecological farming, to protecting and planting forests to sequester CO_2. An entrepreneurial state could generously subsidise farmers wanting to transition to sustainable agriculture, as Germany did when it chose to nurture the growth of renewables.

Such subsidies could be financed through a high tax on pesticides, artificial fertiliser and other harmful farming practices.[68] Farmers who are exposed to the whims of the world market for their income would earn more by producing multiple climate-proof foods. The use of toxins would plummet and biodiversity would benefit.

Instead of dominance and control, collaboration with nature takes centre stage. Alexander von Humboldt would be completely at home in this scenario. He knew that collaboration works. Protection of existing nature reserves and better soil management ensure that the carbon that's naturally absorbed by these ecosystems remains safely out of the air.[69] A third (!) of the CO_2 that has to disappear from the atmosphere in the twenty-first century in order to keep global warming under 1.5 degrees can be sequestered through mass reforestation.[70]

Such projects require space: if one were to plant trees all over the UK, that would take a 'mere' 10.3 billion tonnes of CO_2 out of the air – the equivalent of China's annual emissions. Reforestation therefore doesn't provide a definitive solution while emissions are high. But it can limit the damage. And compared with biofuel plantations, forests are *forty times as good* at capturing and storing CO_2.[71]

The best thing you can do with a forest on a hotter earth? Just leave it be.

In the city, trees can provide relief from localised heat: a street with plenty of trees is 4 degrees cooler in summer than one that's all concrete.[72] 'Living' roofs are also a great innovation: on such a roof you plant flowers and plants in a shallow layer of earth. What you get back is spectacularly multifaceted: a layer of insulation that provides warmth in winter and cooling in summer, filtering and storage of rainwater, a habitat for insects and birds, storage of CO_2 *and* a longer life than a normal roof.[73] The initial costs are slightly higher, but the benefits ... now I'm starting to sound like a broken record.

Greater well-being. The air will be much cleaner in this future. The sources of CO_2 emissions will often also be the sources of air pollution: power stations, factories, cars, buses and aeroplanes.

In 2015, 9 million people died of lung and vascular diseases as a result of polluted air.[74] The death rates from tuberculosis, HIV and malaria put together don't come close to this. Even smoking costs the world less healthy life years than air pollution.[75] Cutting back on fossil-fuelled transport and industry has immediate effects: the sudden drop in air pollution due to the corona-crisis produced tangible health benefits, leading to an estimated 11,000 avoided deaths from air pollution in Europe alone – a bitter fact during a pandemic. 'What if we had this sort of air quality not because everyone is forced to sit at home but because we managed the shift to clean transport and energy?', one researcher wondered.[76] It's particularly people in poorer countries who will reap the benefits of a transition: they're the ones who currently breathe the dirtiest air.

And you know what? We'd also be happier in this future. There's a mountain of evidence showing that people are fundamentally 'biophilic'.[77] When we spend more time in natural surroundings, we feel better. Just the sight of trees reduces stress.[78]

No green paradise, but resilience

If this scenario sounds unlikely, that's because it is. Achieving this picture, or something resembling it, is *much* more difficult than the first scenario, in which we carry on with business as usual. Turning away from the course towards an uninhabitable earth requires mass collective efforts and raises dilemmas that we have never had to face before.

One of the dilemmas involves the resources needed to get a renewable energy supply off the ground. In particular it's a question of metals such as lithium and iron ore for the production of steel. These materials are needed to manufacture wind turbines, solar panels, power grids, batteries and related technologies. In principle there are sufficient supplies of the requisite resources, but they do need to be extracted from the earth, which often – but not always

– involves poor working conditions, exhaustion of already scarce clean water supplies and pollution.[79]

It's a fiendish dilemma: in order to achieve a rapid reduction in CO_2, it's currently inevitable that we will continue to damage nature, and that people in poorer countries will work in mines to supply technology to richer countries – who will always be the first to be able to afford the clean technology.

Colonial relationships will therefore continue to exist, even in this scenario. But there's one big difference with respect to the previous scenario: this time we have a plan for ending the exploitation, as almost all materials used in renewable energy technology can already be recycled now.

And even if you set aside recycling, investing in renewable energy is still a giant step forward from fossil energy, because the extraction and use of oil, gas and coal place a much higher burden on water, climate and environment.[80] Moreover, we can at least try to distribute the yields of the mining of resources more fairly than we have done in the past few centuries.

It won't be a green paradise. We'll undoubtedly make mistakes in this transformation. Every solution creates new problems, enough to keep us busy for centuries to come. But the sooner we make the turn, the sooner we spare the natural world, and the greater the chance that we effectively withstand the crises of climate and biodiversity.

On the way, we'll continue to learn to deal better with changes in our environment and the warming of our habitat, which are already inevitable. To conclude this chapter I'd like to show how switching to sustainability also feeds into resilience.

*

In September 2017, just after Hurricane Irma had left a trail of destruction through the Caribbean and through Florida, another hurricane welled up above the Atlantic. With wind speeds of 280 kilometres per hour, 'Maria' raced towards Puerto Rico. There, thousands of buildings were partially or entirely destroyed, roads became unpassable, the power supply failed. Thousands of people lost their lives. Afterwards, climate scientists noted that both Irma and Maria had caused more damage as a result of a warming climate.[81]

The Canadian journalist Naomi Klein travelled to Puerto Rico three months later. Almost half the inhabitants, more than 1.5 million people, were still without electricity. It wasn't just the power grid that was still gasping; the import of fossil fuels, with which the island produced 98 per cent of its energy, was slow to get going again.

But all that time, and even immediately after the storm, there was a place where the electricity kept on working, where people were able to charge their phones and the elderly could plug in their oxygen masks. In Casa Pueblo, the villa of the local environmental organisation of the same name, the lights stayed on. Solar panels had been installed on the roof twenty years previously. They survived the storm.[82]

*

Some people on the island dubbed Maria 'our teacher', because after the storm people noticed not only what didn't work – 'pretty much everything', Klein writes. They also discovered what *did* continue to work.

Besides Casa Pueblo, there was a second spot that withstood the storm miraculously well: an agroecological farm, high up in the island's mountains. Klein notes how strange it felt to drive for an hour and a half through communities torn to pieces and end up on a farm where laughing children are harvesting beans. Yet that's what she saw.

The farm is managed by the agronomist Dalma Cartagena. She wants to teach local children agroecological principles and familiarise them with the production of their food. A couple of days after the hurricane, Cartagena and the children were already harvesting vegetables again, making compost and thinking about improvements to the farm. The greenhouse and the bananas had perished, but all the vegetables growing close to or under the ground had survived the storm. The students took tomatoes, sweet potatoes and carrots home with them. They were conscious of the importance of their work. 'I feel as if we are throwing a rope to humanity,' a 13-year-old student told Klein.

In the small network of agroecological farmers in Puerto Rico, similar scenes played out. 'Teacher' Maria had shown them how well agroecological farming worked. The diverse crops, the robust roots and the strong soil made these farms more resilient. The majority of the monocultures of bananas, papayas, coffee beans and maize, by contrast, looked like they'd been razed to the ground during the storm.

The story of Maria shows that smart farming and clean energy can increase our resilience. In this future we'll make mass use of these options. When citizens set up these kinds of projects and share the yields, it has the opposite effect of bunkers, gated communities and private islands – it creates communities. On a hotter earth, 'community' is more than just a buzzword. It's a prerequisite for survival.

PART III

WHAT CAN WE DO?

Chapter Seven

ALTERNATIVES AND OPPORTUNITIES

Bees are good at making collective decisions. In spring, after the birth of a new queen, two thirds of the colony break away to find a new home. From a temporary location, hundreds of female bees fly out one by one to explore and discover what would be the most suitable new spot for them.

Hollow trees are always a favourite, but which hollow tree is the best? Choosing a site close to the ground or out in the open can be dangerous.[1] At times a conservative choice is best – a nearby oak, for example – at others an ambitious trek to an elm a couple of miles away.

When an explorer bee comes across a good site, she'll fly back to the swarm and do a dance to promote 'her' location. She'll wiggle her abdomen in the direction of that place and communicate the distance via the duration of her dance. Subsequently, other explorers will fly to that location, and if they too like it, they'll join the first explorer's dance. Each dance symbolises another vision of the future.[2]

The bees abide by a simple rule: the best place merits the best dancing – that's to say, the one performed most consistently and repeated the most times. In this way they boost interest in a particular site and build consensus.

Within a couple of hours or days, a majority of the explorers will be performing the same dance. At that point it's clear which destination has prevailed. In the words of journalist Roy Scranton: 'The swarm has made its decision and takes flight.'[3]

*

Compared to bees, humans are hopelessly disorganised creatures. In the previous two chapters, I outlined two future scenarios, but we'll never make a clear-cut decision in favour of one or the other. Our collective decision-making is a mess. To highlight but one clear difference: our decisions are never free from prejudice and self-interest.

When one of the bees has found a mediocre location, she won't do her best to 'promote' it. That wouldn't be in the colony's best interest. Humans, on the other hand, sometimes champion a future that serves them well personally, but that's disastrous for the collective. In recent decades, oil and gas companies have paid lobbyists billions of euros to influence policymakers. There wouldn't have been any need to do so if they genuinely believed that 'the best destination' meshed with their business model.

In our society, rusty ideas and vested interests have a seat at the table when our direction is up for discussion, making it impossible for us as a collective to make a clear choice for one or the other future.

So we'll never find ourselves in either the first scenario or the second. I outlined the two extremes to show that we really do have a choice – that we can end up in entirely different futures depending on whether we do or don't radically change course. But things won't be as clear-cut as scenario one or two: the forces that propel us towards these different futures exist side by side.

At times that's infuriating. We live in a schizophrenic world in which many governments claim to want to stop the warming yet continue to lend their support to the fossil fuel industry. In one and the same newspaper you can read an article about a new campaign for sustainable living, while underneath it is an advert for ridiculously cheap cruise holidays – cruises being the most polluting holiday imaginable.[4] It's as if we're speaking with two voices, want everything, yet refuse to take chances or make concessions.

At the same time, these contrasting visions of the future also facilitate progress. It's good for us to disagree on what we consider to be our best future. It's only by sharing ideas and knowledge that we find out what works and what's best avoided.

Climate change in the coming centuries

Our emissions determine the eventual rises in temperature and sea level

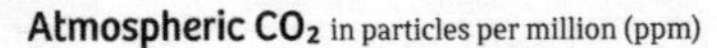

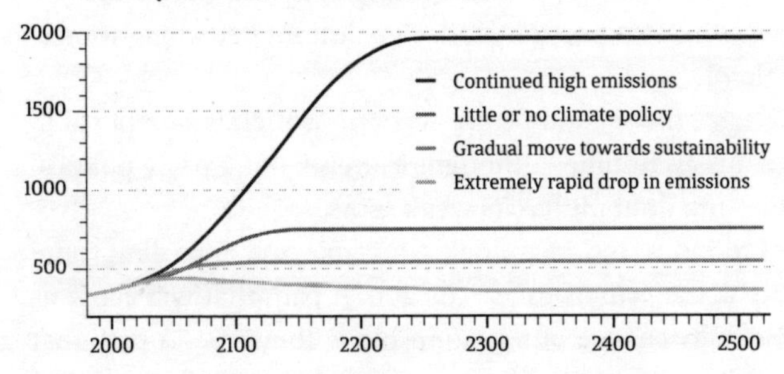

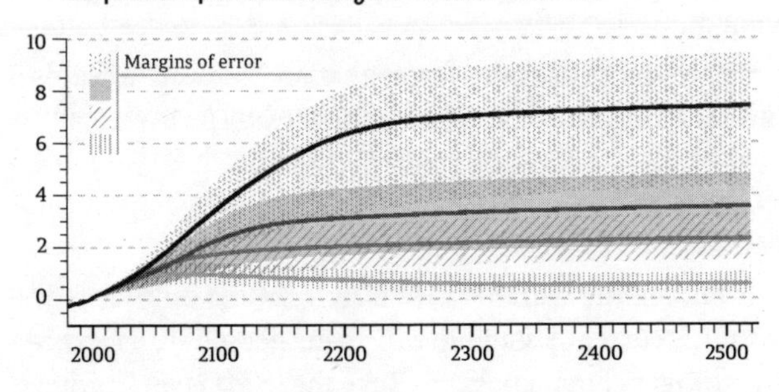

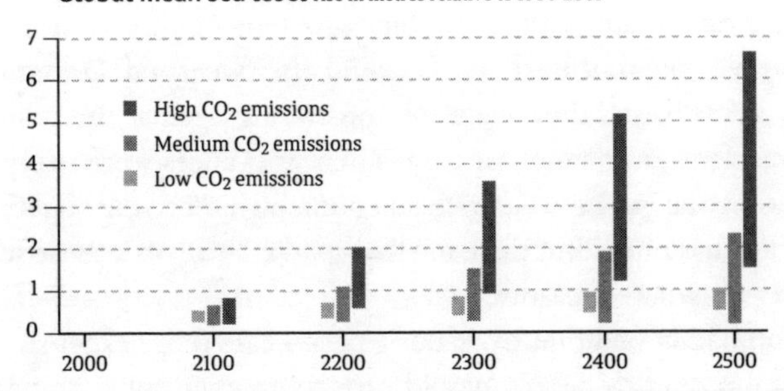

Source: IPCC, *Climate Change 2014: Synthesis Report* (2014), p. 74

An enterprising government that's fully committed to a green future could be a great thing. But a state that's *too* strong could undermine our democratic freedoms. It's only by continuing the public debate, by looking at and discussing opposing visions that we can prevent, wherever possible, today's solutions becoming tomorrow's problems.

In this light, the second future scenario is merely one possibility. Besides, it's an outline, a thought exercise, not a party-political programme – no doubt it has its weak spots.

What I'm interested in is finding a viable and appealing route that will break the grip of the status quo. A path that will allow us to fundamentally change our relationship with nature. A path that enables us to reorganise our society so it no longer leaves a trail of destruction through the web of life. A path with both uncertainties *and* opportunities.

How can we strengthen the forces that are working towards a liveable future? How can we accelerate the Great Turn?

How politicians undermine their own policy

One way of finding the path forward is by working out where the transformation is currently stalling, why many people are *not* exactly rooting for stronger climate policies. Take the '*gilets jaunes*' protests in France.

Kicking off in late 2018, the rallies saw tens of thousands of French people take to the streets weekend after weekend. Dressed in yellow safety vests, they were demonstrating against the policies of President Emmanuel Macron. Their grievances were many, encompassing the gap between rich and poor, high rents, the quality of education and more. But the issue that sparked it all off was a rise in fuel tax – a climate measure.

Macron had hoped that by making petrol cars more expensive to drive, more French people would turn to greener public transport. But this went down the wrong way with those less well off. That big-wig in his palace is worried about the end of the world, the protestors said, while we often don't even know if we'll make it to the end of the month. And public transport doesn't get you anywhere in France, so it was simply impractical too.

All this time, Macron failed to convincingly explain why it was a good measure for both country and climate. Bowing to the immense pressure, he eventually revoked the fuel tax rise.

So what went wrong here?

You might say that Macron introduced the wrong measure. In the light of a tax reduction for higher-income earners he'd introduced earlier on in his presidency, this climate policy wasn't fair.[5]

Nor was it a good idea to hit middle-income earners who didn't have a good alternative. Macron should have given them something tangible to win their support. In the Canadian province of British Columbia, for instance, there's a great deal of support for a CO_2 tax, as the proceeds are shared equally among all citizens.[6] Pollution is taxed across the board and everybody benefits from it. A similar rebate for the French might have worked wonders.

But the problem is deeper. Look at the debate in the US about the Green New Deal (GND), an ambitious package of policy measures including a switch to fully sustainable electricity within ten years (!). As soon as progressive democrats started talking up the plan in late 2018, a fierce debate ensued. Conservative media like Fox News aired segment after segment about it – devoting much more time to the plan, in fact, than more moderate or left-leaning channels like MSNBC and CNN.

The main message from conservative politicians and talking heads was that the left was planning to destroy the economy with an 'impossible and unaffordable' plan.[7] They wanted to 'take away your hamburgers'.[8] Within months, 80 per cent of Republican voters strongly opposed the Green New Deal.[9]

The only thing many progressives and moderates could do was deny the charges. No, in fact, the GND didn't plan to 'ban air travel and cows', as President Trump had tweeted. But the terms of the debate were set: for many Americans, this was about what they would be losing.

This is a recurrent theme in other countries' climate debates: confronted with criticism that climate plans would be too sweeping or costly, moderates begin to talk down the implications. Anything to avoid the suggestion that voters might be negatively affected. Everything will stay the same, except we'll cut emissions – that's the moderate message on climate change.

But here's the thing: we urgently need to talk about *where we're going*. The prospect of 'fewer emissions' is meaningless in and of itself. Why should we opt for a future without CO_2 emissions? Will there be coffee on arrival? Will we have fun along the way? Will the transition create jobs? Will it increase or lessen the burden on low-income households? How can we make sure that the fight against global warming doesn't widen the gap between rich and poor?

There's not much point in having a climate policy if it's not informed by a vision, by a story about what we stand to gain from it. Right now, the message we keep hearing most is that something's going to disappear. Not because it's fun, but because it's necessary, because otherwise we have no future.

That's not an attractive prospect – it's despair.

A shared story

Only a compelling story can help us out of this deadlock. Man as 'master and possessor' of nature enjoying amazing triumphs and victories – that was a great narrative. It brought immense benefits to large parts of the global population and it established man's place on earth. For many, a growing economy meant progress and freedom and comfort. That's what it was all about.

Now that this narrative is collapsing like a house of cards, we need a new one. Giving people nightmares about the 'climate apocalypse' isn't enough – it won't prevent disasters.[10] IPCC reports won't do the trick either – hardly anyone ever reads them, and they're not exactly inspiring.

At the end of the day you can only replace a story with another story.[11] What kind of world would we like to live in? What makes life worth living?

These were the questions that Puerto Ricans had to ask

themselves after Hurricane Maria wreaked havoc on their island in 2017. All kinds of ideas about what constituted a good 'recovery' were circulating at the time. Stronger market forces, more tourism, more foreign investors, or something entirely different?

'We need to know where are we heading,' a local environmental activist told Naomi Klein. 'We need to know where is our ultimate goal. We need to know what paradise looks like.'[12]

What does our paradise look like? The more people have a shared view of our destination, the better we can organise our efforts accordingly.[13] An appealing future narrative for us is the equivalent of an appealing dance for a colony of bees.

To me – surprise, surprise – a green transformation, some kind of variation on the second future scenario, is the most attractive option. It provides a foundation for a positive story about progress in the face of setbacks, about people cooperating and rising above themselves. It has the potential to become an epic story filled with action, determination and transformation – a great deal more exciting than the straitjacket of the first future scenario, the soul-destroying lack of alternatives in that world.

And the best thing is: all the ingredients for a convincing story are already in place. Most of them have nothing to do with politics. You don't have to identify with the left or the right to care about clean air, or about seas in which tropical coral reefs and the wondrous creatures they produce have a future.

If you want the animals from children's stories and nature documentaries to survive, then the appeal of the green future narrative is obvious.

If you're against further European integration, or you just dropped out of the EU, you can see renewable energy as a means to achieving greater national autonomy – after all, each wind farm and each hydrogen plant breeds a bit more independence from foreign oil, gas and coal.

If you have the uneasy feeling that a globalised world economy is out of touch with humanity, then a sustainable future offers better prospects: looking out for one another and taking care of the land we live on is the guiding principle here.

If the idea of mass migration is a concern, then an effective climate policy is a good measure against the crises that trigger such large-scale exoduses.

And so there are a million reasons for opting for the green future narrative.

Of course not everybody will reap the benefits in the short term. There's a lot at stake for people whose work, identity and fortune are inextricably linked to the fossil fuel system. If you've spent years putting your heart and soul into finding new oil reserves, a green future will rather take the shine off your efforts.

And if you're working in a garage, servicing or repairing cars – as 1.5 million people in Europe do[14] – then you're not necessarily going to be thrilled about the fact that electric cars require far less maintenance because they have far fewer moving parts.

And those who've bought a bunker or an apartment on a private island will want persuasive arguments to change tack and participate in the Great Turn.

Is it possible for the story about a green future to ultimately become an attractive proposition for *everyone*?

I think so. Putting aside all political, moral and ecological considerations, what's left is a rock-solid argument. The economy. Analyses show that a transformation towards true sustainability will lead to millions of new jobs, lower healthcare costs, less climate damage and a rise in incomes.[15]

A recent report estimates that the UK, for instance, needs to spend up to £20 billion a year until 2050 to meet its climate targets. That's less than 1 per cent of its current gross domestic product (GDP).[16] It's around one tenth of what the UK spends every year on healthcare. Even if it were twice as much, at £40 billion a year climate spending would still be less than what UK residents fork out yearly on holidays (£46.5 billion in 2018).[17]

What it boils down to is that Western countries can easily afford to make their economy more sustainable.

The economic downturn after Covid-19 does not refute this fact, but underscores it. In no time at all, rich countries were able to shell out hundreds of billions of dollars, euros and yuans to ensure businesses would survive the crisis. Then, when recovery packages were drafted, it soon became clear that investing in climate solutions would offer the best returns for government spending.[18] Improving energy efficiency by pouring money into insulation of the building stock, for example, was identified as a win-win measure, since it would simultaneously 'support existing workforces, create new jobs and drive reductions in emissions.'[19]

And there are other strong economic incentives to do so. Joining the green switch late or only partially carries its own risks. The spread of renewable energy will force the established fossil fuel economy's hand. History has taught us that an existing sector will start to shrink shortly after a viable alternative appears on the market.

Nearly all film roll manufacturers went bust when the digital camera appeared on the scene. The problem wasn't so much that everybody stopped buying film overnight, but that the Kodaks of this world had been counting on growth, which was now suddenly guzzled up by the digital competition.

So established businesses aren't just hit once a rival dominates the entire market. The key question is *who gets the growth?* In the field of energy, the answer to that is now more and more often: the sustainable alternative. In 2018, renewables made up almost 70 per cent of all newly added electricity generating capacity in the world.[20] Because the prices of solar panels and wind turbines continue to drop, these renewables will be meeting ever more of the demand in years to come, at lower costs.

That's one of the reasons why in the US more coal-fired power plants are closing down under Trump than under Obama, despite Trump's pledge to revive the coal sector.[21] In the next few years, established energy companies such as Shell and BP will be finding it harder to keep growing in their old oil and gas market. Their green competitors are still relatively small, but they're big enough to inflict some real damage.[22]

Investors are already seeing a payoff in dumping fossil fuel.[23] An analysis from October 2018 showed that the New York State Common Retirement Fund, a civil servant pension fund, would have earned an extra $22 billion had it divested its shares in fossil fuel companies ten years ago. Such a move would have yielded $19,820 per pension holder.[24] British public pension funds have lost nearly $1 billion following the decline of coal, another analysis has shown.[25] In the same vein, BlackRock, the world's largest asset manager, would have served its customers much better had it divested from fossil fuel companies ten years ago.[26]

Another attractive benefit of switching? A sustainable energy system is far more efficient. Or perhaps it would be more accurate to say: the fossil fuel system is excruciatingly wasteful.[27] Two thirds of the fossil fuel we're using now is lost as useless waste heat. For instance, in order to generate electricity from coal we must first bring water to the boil and use the resulting steam to run a turbine. By running the turbine we generate electricity, but approximately *half* of the energy from the coal is lost in the process.

Such losses are everywhere in the current energy system. Three quarters of the petrol you burn in a car is lost to internal friction in the engine and to warming up the engine block. That doesn't get you anywhere.

Sustainable alternatives can offer a genuine improvement. Electric cars, for example, are nearly four times more economical than their petrol or diesel counterparts. A heat pump uses a quarter of the energy of a natural gas boiler. And we'd need fewer large, inefficient power stations if solar panels and wind turbines were to generate our electricity – at most we'd use them to burn hydrogen. That's not to say that the production of such 'solar fuels' doesn't also incur major losses – it does. But on the whole, an economy that has switched to green energy will be far more efficient – and the energy is likely to be cheaper too.[28]

Against the background of an economy that's under pressure from global warming, green investments are a no-brainer. It's simply much cheaper to tackle climate change than to allow it to happen.[29]

Will depletion run its course?

Call me naïve, but for a long time I hoped that the arguments outlined above would automatically lead to change. Maybe they haven't been pieced together into a thrilling story often enough – or maybe this story hasn't been told compellingly enough.

But the arguments aren't new. To give you an example, the authoritative, 700-page *Stern Review on the Economics of Climate Change* from 2006 calculated that a green switch would cost the world around 1 per cent of its combined GDP annually. But according to the same report, climate change damage would amount to an annual loss of at least 5 per cent of this total GDP, from now on and for all eternity.[30]

That's an argument in favour of action if ever there was one.[31] But until now, fears for the losses incurred by specific sectors such as aviation have outweighed such figures. The grip of the status quo is too strong, the schizophrenia I mentioned earlier too entrenched. Escaping it is going to take a huge amount of inventiveness and perseverance.

That's the conclusion I've reluctantly arrived at in recent years. A trip to Ecuador really opened my eyes.

In the spring of 2017 a colleague and I travelled to Ecuador as part of an exchange with local journalists. While based in the capital Quito I worked on stories about oil extraction in the Amazon. In exchange for loans from China, socialist president Rafael Correa had doubled down on oil exports from his South American nation. Even previously protected areas of the rainforest had been opened up to mining and oil extraction. What was happening in Yasuní National Park has stayed with me as a textbook example of what you get when you let the global economy take its course.

Yasuní is in the east of Ecuador, in the Amazon. It's a protected area because the park is home to more species of birds than the whole of Europe, more species of toads and frogs than the whole of North America, and more insect species than anywhere else in the world.

The park has it all: rare medicinal plants, endangered apes with colourful names, such as the white-bellied spider monkey, as well as pumas, jaguars, ocelots, gigantic anteaters, leopards and sea cows. Also living in the park, in voluntary isolation from the rest of the world, are several thousand members of the Waorani tribe.

In 2007, an oil field with a volume of more than 800 million barrels was discovered underneath the park. Enough to supply the global economy with ten days' worth of fuel – and enough to destroy Yasuní. Leave it in the ground, you'd think.

But Ecuador didn't have that luxury – the country needed money and it had promised China oil. The new discovery boosted the country's recoverable oil reserves by 20 per cent. A conservative estimate put Ecuador's potential earnings from the fuel at $5.7 billion, at a time when half the population was surviving on less than two dollars a day, and there was huge international demand for oil. Prices were high; it was almost irresistible.

But Correa was under immense pressure from conservationists and indigenous tribes, who – like a vast majority of the population – were vehemently opposed to the drilling. A cunning plan was hatched. Ecuador was going to try to collect money from the international community in exchange for leaving the oil in the ground.

The plan would see other countries help pay for Yasuní's preservation, leaving Correa with enough money to develop his country. He launched a fund and had a whip-round on the international stage to raise $3.6 billion, 60 per cent of what Ecuador could expect to earn from drilling at the national park.

It was a good idea. It made the international community jointly responsible for this stretch of nature. If enough money was raised, the park would be closed to oil drilling once and for all, and over half a billion tonnes of CO_2 wouldn't be released into the atmosphere.

But the plan failed. The world didn't understand what Ecuador was asking, saw it as blackmail, or simply wasn't prepared to cough up. Only $300 million was pledged, of which a paltry $13 million was actually transferred.

In 2013, Correa gave the go-ahead for oil extraction in the park. Three years later, 23,000 barrels of oil a day were being pumped out.

Meanwhile, roads have been constructed in the park, which marks the beginning of the end for a nature reserve. Where there are roads, there are people, and where there are people, there is, at least in the current system, deforestation. Like so many drilling sites in the land, the area where the oil is being extracted has been militarised. Up until the end of his term in 2017, President Correa was something of a dictator who dismissed those against the drilling as 'opponents of progress'. He ruled over the oil extraction with an iron fist. He had no choice. Once you're dependent on oil exports, there's no going back.[32]

It was in Ecuador when the penny finally dropped, when it became clear to me what we're doing. And it wasn't in Yasuní, because I never went there myself. It was in the massive garden of Heike Brieschke, a German ornithologist (a bird expert), who has worked in Ecuador for more than a decade. On her 11-hectare piece of land she has counted no fewer than 220 bird species, including twenty-eight kinds of hummingbird that hum and whir as they hover in front of a flower to drink the nectar.

A fellow journalist took us there. While all the hummingbirds I saw had long, pointy beaks, their shapes and colours were totally different. One specimen had white feathers around its feet, looking as if it permanently wore cute little slippers. As if evolution was poking fun at it.

While serving us herbal tea, Heike told us about the training centre on her land, from where she's trying to raise the environmental consciousness of local kids. The government wasn't really doing much about this, she said with a worried frown. By now, I'd learned that Correa's government was thinking of designating even these kinds of bountiful places as oil-drilling sites.

Amidst all of these shades of green, it finally dawned on me what a shame that would be. And I understood what people mean when they talk about 'natural wealth'. Natural wealth is this: more shades of green than you can possibly count and a hummingbird wearing slippers.

Back in the Netherlands, I saw a grainy photo of a jaguar that local journalists claimed had been taken in Yasuní.[33] Majestic. Padding along a stack of pipelines. Absurd.

Oil extraction in the Amazon shows how reckless our economic system is – how far it forces us to go, and that just about everyone is complicit. The plane that took me to Quito and back to Amsterdam again was filled with the kerosene that's produced at sites at like this. Me 'opposing' it doesn't change a thing.

A 'tragedy of the commons' is the term American ecologist Garrett Hardin used to describe the bind we're in. We're all using scarce common 'goods' – oil, gas, coal, air, soil life, a liveable climate – but hardly any of us are presented with the bill for the depletion or destruction of them.

If I cause a tonne of CO_2 emissions, I'm not forced to pay for the consequences, such as the 3 cubic metres of Arctic sea ice that melt as a result.[34] When a farmer uses so many toxins that he effectively wipes out all the multicellular organisms on his land, it will only cost him a drop in yield – whereas the whole world will ultimately pay for the decline in biodiversity.

The British investor and asset manager Jeremy Grantham didn't mince his words in 2018 when he said, 'We deforest the land, we degrade our soils, we pollute and overuse our water, and we treat our air like an open sewer. All of this is off the balance sheet and off the income statement. Capitalism and mainstream economics simply cannot deal with these problems.'[35]

Capitalism's main pretence is that the market always accurately reflects the value of goods. Thanks to the pure laws of supply and demand, so the theory goes, we always pay the right price for what we consume.

Turns out that's complete rubbish. Nobody is being charged for gradually making the planet unliveable. The tragedy is that what belongs to everyone ultimately belongs to no one – and then disappears.

Yet none of this is really getting through to our politicians. Those in the centre ground, the right-wing parties and the established fossil fuel players, especially, still adhere to the belief that 'conservative' climate policy is the most viable and also the smartest route. They pass this off as the position of realists, of sensible men and women who are keeping their wits about them, who actively resist the so-called 'alarmists' who want too much change too fast. They're keeping a cool head, even on a warm planet. But these self-proclaimed realists are sleepwalking into an era of enormous economic damage, with repercussions for us all.[36]

As I said: the forces that are steering us towards the first future scenario exist side by side with those working towards the second. You can think of the corona-outbreak as a very short pause on these opposing forces; soon they were wrestling again – over bailout money and recovery packages. And history shows us they don't give up that easily. Take the lobby in the European Union.

A recent tally reveals that between November 2014 and August 2017 gas industry lobbyists had a total of 460 meetings with the two commissioners responsible for the EU's climate and energy policy. Representatives of green NGOs only got as far as fifty-one meetings with the pair.

This means that the European top officials who decide on our energy and climate policy have *nine times as many* meetings with the fossil fuel industry as they do with environmental advocates.[37] The gas sector is using this access to European leaders to get new projects off the ground. It's no wonder then that Europe claims it wants to become greener, yet keeps pushing for the development of new natural gas infrastructure. Europe now risks spending €29 billion on unnecessary gas projects.[38]

Will things run their course? Will the depletion of the earth's resources reach a limit at some point?

Don't count on it. Despite competition from renewables, oil and gas companies are doing absolutely everything they can to extract ever more fossil fuels. The likes of Shell and ExxonMobil are now using solar panels to generate the energy needed to extract oil and gas from the ground.[39] That's a bit of a paradox, I hear you say. Absolutely. Google, Microsoft and Amazon are providing the computing power and the algorithms to make the oil drilling of the future more efficient.[40]

As long as the oil and gas companies remain profitable, and as long as governments keep accommodating them, they will plough on. There's enough oil in the ground to maintain our current levels of consumption for a further 227 years. Natural gas will last us for another 270 years or so, or even more than 1,500 years if we decide to extract it from the frozen bottom of the ocean – a technique that Japan, China and the US are currently investing in.[41, 42]

I struggle to get my head round the idea that someone would deliberately set out to develop such a new extraction technique; it's what you'd do if you actively wanted a hothouse earth.

But that's the bizarre reality we now live in: it seems inevitable that the death knell has been sounded for oil and gas extraction, and it seems equally inevitable that this extraction will nevertheless continue.

Allowing the ongoing emission of ever more greenhouse gases is extreme. And yet many leaders are acting as if it's extreme to want to change course much faster.

'How could we deem "realistic" a project of modernization that has "forgotten" for two centuries to anticipate the reactions of the terraqueous globe to human actions?' asks French philosopher Bruno Latour. 'How could we accept as "objective" economic theories that are incapable of integrating into their calculations the scarcity of resources whose exhaustion it had been their mission to predict? [...] How could we call "rationalist" an ideal of civilization guilty of a forecasting error so massive that it prevents parents from leaving an inhabited world to their children?'[43]

Only governments can halt this downward spiral – they're in a position to rewrite the rules of the game. But governments are not or not sufficiently rising to the challenge. Competitive edge, short-term employment and growth occupy the minds of our leaders, especially in the light of the corona-crisis and the mountain of debt it left us with. Already, there is talk of austerity – the exact opposite of the much needed green stimulus. Already, states are trying to boost extraction, production and consumption – doubling down on the plundering, repeating mistakes of the past. As the biologist Bill Mollison concluded, 'Evil is rigorously applied stupidity.'

There's only one possible conclusion. A new narrative of the future will have to encompass more than a 'yes' to the alternatives. It will also have to include a loud and clear 'no' to continuing down our current, hazardous route.

In the final two chapters, I'm going to introduce you to the global movement of people who are taking matters into their own hands and are fighting for a liveable future. They are doing so by drawing clear boundaries and by working on a sustainable economy.

Chapter Eight

THE BATTLE OF THE CENTURY

All over the world, people are hitting the brakes. They will no longer allow the fossil fuel system to charge on as if there were no problem.

This chapter is about those people. About the naysayers, those taking a stand, the fighters on the front lines of the climate debate. Here I'll describe three places where they're campaigning: in the courts, among investors and around the infrastructure of the fossil energy supply.

In all three campaigns, the arrows are aimed at oil, gas and coal; and rightly so – by causing more than three quarters of all emissions, fossil fuels present the biggest challenge for the climate. Experts agree that reducing these emissions should have the highest priority, given that there are excellent alternatives for oil, gas and coal.

Other sectors, such as the food supply, where emissions are harder to eliminate, will then have more time for transformation. That's also fair: the alternative to a cow that emits methane is no cow – for that you have to persuade people to stop eating meat and dairy products. The alternative to a coal power station is a wind park with a hydrogen plant, a much more achievable step.

Let me begin by eliminating a much-heard caricature of activists: none of the people I'll introduce you to in this chapter is claiming that we can go from using fossil fuel to doing without it overnight. That's not what they're fighting for. What activists want is to escape the grip of the current system. They're identifying the weak links in the chain from borehole to petrol pump and inserting their crowbars there.

You could say they're waging the battle of the century.

Lawyers in action

The first strategic campaign is taking place in the courts.

A court isn't the most obvious place to look for a breakthrough on the climate front, I admit. But as I travel to the Palace of Justice in The Hague on 9 October 2018, I'm still hoping to witness it. Today the court is passing judgement on the higher appeal by the Dutch state in what has become a famous – and infamous – climate case with globe-spanning repercussions.

The action group Urgenda started this case against the Dutch state back in 2013. And to many people's surprise, in 2015 Urgenda got precisely what they demanded. The court tasked the state with reducing emissions by a minimum of 25 per cent by 2020 compared with the level in 1990. Anything less would be 'negligent', because the government should protect its citizens against the threat of climate change. The cabinet, like many commentators, felt that the court had overstepped the mark, and immediately after the ruling, it launched an appeal.

So that's why we're in court again now – Urgenda's lawyers, the state's lawyers, dozens of journalists and onlookers. Will the 2015 ruling be upheld? When the judge starts on her verdict, the people around me hold their breath. For the next twenty minutes we're witness to a reckoning.

The court makes mincemeat of *all* the arguments the government lawyer put forward to exonerate the state – that the Netherlands is already doing enough, that the Paris Agreement will ensure a reduction in emissions, that the climate targets are the business of politics, not the court. Not a single point is upheld.

The Dutch state must do more to protect its citizens against climate change, says the cast-iron ruling. A lax climate policy puts people at risk, the court confirms, and there is a limit to the freedom of politics to endanger people. That limit has been reached.

The fact that other countries are doing too little does not absolve the Netherlands of its responsibility, the judge also notes. The Netherlands' contribution to the global problem – approximately 0.3 per cent of annual emissions – is big enough to be relevant.[1] The state will again be tasked with reducing emissions faster than was attempted in the existing climate policy.[2]

The ruling is transmitted across the world at lightning speed, as it was three years ago. The verdict in the Dutch climate case invigorates dozens of lawyers who are working to compel more climate action from governments in their own countries. In 2015, these lawyers around the world took inspiration from the Urgenda ruling; now they feel all the more how powerful the law can be as a means of breaking open political deadlocks and ingrained interests.

*

The Maastricht litigator Roger Cox is one of the instigators of this worldwide judicial movement. In recent years he has experienced the way 'ripples caused by citizens can lead to a powerful wave'. Everyone can 'be part of the global social movement of committed citizens', says Cox.[3]

Cox hasn't always been such a revolutionary, as I've discovered during conversations with him in recent years. Until 2006, he worked with disputes in the construction and property sector. He had no concerns about the changing climate. Minus 50 or minus 49 at the North Pole, he thought, what does it matter? It's still crazily cold.

Until he saw *An Inconvenient Truth*, the climate documentary presented by Al Gore, in 2006. It's really an alarming PowerPoint presentation, with Gore climbing onto a cherry-picker to indicate how ridiculously high our CO_2 emissions are compared with the natural variations in the past 600,000 years.

Cox saw the film with his wife Saskia and was utterly shocked. Why didn't he know this? Surely this was what the government was for? But they had barely mentioned the climate action necessary. Cox felt betrayed.

From that moment on, he became more and more preoccupied with it. He worked with his wife to organise showings of Gore's film throughout the entire country. He spent evenings and weekends reading IPCC reports about previous ice ages and melting ice caps. He lobbied in his province, Limburg, for a 'circular economy' in which every material would be recycled so that waste and emissions disappear. He founded a certifying institute with the creators of the design philosophy Cradle to Cradle, who have the same aims.

Whether his efforts will change anything, he doesn't know. But he can't stand the idea of the earth continuing to grow warmer and his daughters later asking, 'Well Dad, didn't you know that?' and having to say, 'Yes, I knew it all, but what was I supposed to do about it?'

In subsequent years, Cox effectively had two jobs. In the daytime he was a lawyer for the construction sector; in the evenings he was a climate activist. In 2010, that changed. He realised that the civil law he worked on in his day job also offered good opportunities to take action for the climate. Suddenly the penny dropped: those allowing climate change to happen were creating a *dangerous situation*. And that was unacceptable. Dutch judges have been loud and clear on this matter.

The most famous example is the case known as the 'cellar hatch ruling' of 1965, in which the Supreme Court judged that a Coca-Cola employee who had left the cellar hatch of an Amsterdam café open was responsible for the injury of a visitor who fell into the hatch. The drink supplier should have prevented the injury by warning the person about the open hatch, or putting a crate in front of it.

Cox realised global warming is effectively a huge cellar hatch story. Greenhouse gas emissions constitute risk. States can pursue a stricter climate policy and companies can invest more in sustainable alternatives to eliminate the danger. Together they have the greatest relative power and can therefore exert an influence. But they're not doing it, or in any case not enough, because emissions are still rising.

He started writing a book to organise his thoughts on possible climate cases, concluding that the law can contribute to an 'all-encompassing energy revolution'. Marjan Minnesma, director of Urgenda, read the book, contacted him and brought in another lawyer, Koos van den Berg. The three began to work on a lawsuit, and Minnesma arranged for committed citizens to register as 'co-plaintiffs' and come up with ideas for the complaint.

On 20 November 2013, the state was sued by Urgenda and

almost 900 co-plaintiffs. Minnesma called it a lawsuit 'of mercy', because politics was so 'impotent'.[4] After all, successive cabinets had failed to end the destructive spiral of endless growth at the cost of nature. Cox and Minnesma wanted to force a breakthrough.

In June 2015, the court found in Urgenda's favour. What swung it their way? In the summons, Urgenda appealed to uncontroversial judicial principles. Since the cellar hatch ruling, there have been clear criteria to determine whether someone has been negligent, and thus allowed a risk to arise.

Cox and Van den Berg make a clear case that the state had in fact been negligent. The Dutch government acknowledged the threat as early as 1992, when the Netherlands, along with 190 other countries, signed the first global climate treaty. They expressed the intention of preventing 'dangerous anthropogenic interference with the climate system'.

In climate science and international climate policy in subsequent years, it became a generally accepted conclusion that the wealthiest countries would have to reduce their emissions by at least 25 per cent by 2020 compared with 1990 in order to prevent such 'dangerous anthropogenic interference' (currently defined as 2 degrees of warming). But the Dutch policy, as the court ruled in June 2015, was so far only on track for a reduction of 17 per cent by 2020.

So the state was failing to do what the state itself claimed was necessary to protect citizens. And because it has a duty of care for citizens, the court intervened. It ordered the state to reduce emissions by 25 per cent in 2020.[5]

The verdict proved what some jurists had long been saying: that every democratic state has a legal duty to protect its citizens against climate change. Anyone who sees danger looming and has the opportunity to minimise or prevent it has a legal obligation to do so. Ultimately, it's human rights that underpin this duty. It's one of the most fundamental principles of any judicial system.[6]

After the Dutch state lost its appeal in 2018, it went to the Supreme Court. Again, it lost. Every country must do its share to reduce emissions, the Supreme Court ruled, and 'no single reduction is negligible'.

The relevance of the Urgenda case was immediately apparent to other countries. A few climate lawsuits had previously been filed elsewhere in the world, but until around 2005 they were rarely successful. The Urgenda ruling set a new wave in motion; in 2017 there was a clear peak in the number of new climate cases.[7] From Pakistan to Switzerland, and from New Zealand to the UK, in twenty-eight national courts, it has been argued that governments who fail to bring emissions under control are negligent – and that the court can therefore intervene.[8]

Far from all those cases have been won. But even the climate cases that fail still have an impact, because the government has to account for its actions in a way that generally isn't required in parliament. In a parliament, the majority generally agrees with the ruling party: party discipline means that ministers come away with vague commitments.

In court that doesn't work. Nor does spinning a nice story about it. If the existing policy is insufficient to reduce emissions in line with the international climate agreements, that becomes as clear as day. A lawsuit also draws media attention, which helps to make the dangers of global warming clear to a wider public. A court verdict has a hard quality that's lacking from your average government advisory board. A court ruling brings obligations with it.

All the lawyers working on climate cases around the world are aware that their work is groundbreaking. Their governments tend to oppose them fiercely in the initial phases, but lawyers are known for sticking to their guns. As in the US, where lawyer Julia Olson pursued a case against the federal government on behalf of a group of twenty-one children from ten states. Olson wants to guarantee the 'right to a safe climate' – so she wants the government to stop facilitating extraction of new oil, gas and coal. The White House has pursued the case aggressively and achieved a provisional win in January 2020; the Ninth Circuit Court of Appeals judged that although it was clear that something needed to happen, climate policy was not the business of the court. 'This case is far from over,' Julia Olson responded; she's taking it to the Supreme Court.[9]

So too in Belgium, where a climate case modelled on that of

Urgenda has been delayed by three years, first because of a legal battle over the language in which it would be tried. Meanwhile, though, that case has gathered more than 55,000 co-plaintiffs. The French climate case that began in 2018 can count on the support of more than 2 million people.[10] One suit modelled on Olson's work was launched in Canada in late 2019. One against Heathrow's construction of a third runway succeeded in the UK a few months later – the construction permit was ruled illegal, beacuse it was incompatible with the Paris Agreement. And so on and so forth.

*

Roger Cox has now thrown himself into a new case, this time against Shell. On behalf of Friends of the Earth Netherlands, six other environmental organisations and more than 17,000 private co-plaintiffs, in April 2019 he filed a lawsuit against the seventh-largest oil company in the world. The demand is that Shell wind down its oil and gas production so that it no longer contributes to climate change in 2050.

It's the first time in the world that an environmental organisation has demanded a change of course from an oil company in court.

Shell is currently undermining the aims of the Paris Agreement, Cox states in the summons. The big investments that the company continues to make in fossil infrastructure contribute to the Paris targets becoming unachievable. After all, it is patently Shell's aim to supply *more* gas and oil, while emissions need to fall. The fact that we're racing past the 2-degree limit is not Shell's fault *alone*, Cox states, but the company can be held accountable for its contribution.

If the state can be held accountable, then Shell can too, the environmental organisations argue. It may be that Shell's clients ultimately cause the most emissions – they consume the natural gas and petrol supplied by the company. But since 2017, the company itself has acknowledged that those emissions are Shell's responsibility.[11] A bit like the quality of a can of paint is the responsibility of the manufacturer, not the consumer.

Cox tells me the case revolves around a single question. Can a multinational such as Shell knowingly contribute to the destruction of the human habitat worldwide? Or is this company, too, obliged to be conscientious, like the drinks supplier in the cellar hatch ruling?

If Cox can show that Shell was aware of the risks and could have chosen a different path, he may win.

Shell has called the case 'inappropriate, misleading and legally unwarranted'. In its initial reaction, the company stated that the Paris climate accord applies only to states, not to private companies. Moreover, Shell is already subject to environmental legislation, the company's lawyers write, and CO_2 emissions are not illegal anywhere. In selling products that cause CO_2 emissions, Shell is not breaking any laws.

Cox thinks he has a counterargument to that. He makes a comparison with asbestos, which was used until well into the 1980s, despite the fact that in 1969 there were already enough reports showing it caused cancer. The companies that subsequently continued to work with it were later convicted for its deadly consequences to employees and consumers.

So historically it wouldn't be the first time that companies are convicted despite having followed the rules. No licence on earth can definitively exempt a company from liability – because no licence states that you can cause (climate) damage to others.

A possible conviction for Shell could be a springboard to lawsuits against other oil and gas companies, just as the ruling in the Urgenda case formed a catalyst for dozens of other cases. 'It's a link in a much longer chain,' says Cox.[12]

*

Besides the climate lawsuits about preventing harm, another judicial route is being explored: that of compensation for damages. In the US and elsewhere, lawyers are working on behalf of local authorities to try to recover the costs of necessary climate measures like coastal defences from companies such as ExxonMobil and BP. This is comparable to the way tobacco manufacturers from the

1950s onwards have been held liable for the harm they caused to smokers. For decade after decade, the oil companies have been able to profit from fossil energy. Isn't it then fair that they now bear some of the public costs?

The oil companies, after all, knew perfectly well what damage they risked. Take the example of ExxonMobil. Research by a predecessor of that company, Humble Oil, already showed in 1957 that burning fossil fuels leads to the accumulation of CO_2 in the atmosphere. In 1977 the entire Exxon board saw a presentation by one of *their own* climate researchers about how serious the consequences of the emissions could be.[13] At the time, Exxon was investing large sums in its own climate research: for instance in 1979 it started measuring the CO_2 concentration of the air with instruments installed on an oil tanker (ironic, yes).[14]

But what did the company do with that knowledge? Did it call for action? Did it share its conclusions with the world?

No, Exxon lied almost continually to the public about the dangers known to the company.[15] It intentionally sowed doubt and financed PR campaigns and individual climate scientists who actively challenged the emerging consensus about the risks of global warming.

In 1989, Exxon was one of the founders of the Global Climate Coalition, a lobby group with the aim of undermining belief in the seriousness of climate change. Shell was also a member of the group, and remained in it for almost ten years. This coalition of fossil fuel companies did everything in its power to spread doubt, with advertisements, advertorials in national newspapers, special memos for journalists and by lobbying members of Congress in the US. In part as a result of this, the US refused to ratify the Kyoto climate treaty in 1998 – the very treaty in which it was agreed that rich countries should lead the way in reducing emissions because they had caused the largest share of the problem and had the most resources to tackle it.

Blocking Kyoto was precisely the aim, as we know from a leaked memo from the American Petroleum Institute (API) from April 1998. It contained a 'plan of action': 'Victory will be achieved when

... average citizens "understand" uncertainties in climate science ... [and] recognition of uncertainties becomes part of the "conventional wisdom" ... [and] when those promoting the Kyoto treaty on the basis of extant science appear to be out of touch with reality.'[16]

Those involved have always denied that this plan of action was literally carried out. But the well-known lobbying practices of the fossil energy companies indicate that undermining science is a conscious strategy – particularly in the US. No holds were barred in undermining a policy that was judged to be 'draconian'. And then the campaigns sowing doubt blew over from the US to Europe; 'reservations' financed by fossil energy companies began to colour the debate here, too.

Climate lawyers are now exposing the intentional campaign of doubt in order to support the argument that compensation is warranted. In the US there are fourteen such cases currently running – New York, Baltimore and San Francisco, among others, have initiated proceedings. From the charges it seems that there is ample evidence that the fossil energy companies knew that their activities represented a threat to the climate *and* that they had other options.[17]

For instance, it was the oil companies who were quick to apply for patents on energy technologies that could substantially reduce emissions. In 1967, when the American government planned to invest in electronic alternatives to combustion engines, the chair of the API said to a Senate committee from the US Congress that this was completely unnecessary because his industry was already doing plenty of research on the matter.

The climate cases against the fossil fuel industry in the US have achieved little so far. The federal judges who ruled on the New York, San Francisco and Oakland cases held that climate policy is a matter for Congress and the White House, not the federal court. The battle to see who is responsible for what and in which courts these cases should be heard will now be fought out in the appeals courts.

At any rate, the climate lawsuits compel the parties involved to talk about very specific obligations and responsibilities. What exactly did you know, and what could you have prevented? What was your share and your responsibility?

Meanwhile those questions are being put not only to countries and energy companies, but also to insurers, banks and investors who facilitate the continued extraction and combustion of fossil energy. Individual citizens can be certain that they will never be sued for their complicity in global warming: the contribution of a single person among a total of 7.8 billion world citizens is negligible from a legal – if not moral – perspective.

But countries, large organisations and perhaps even chairs of boards *can* be questioned about their contribution. Accountants, jurists, investors, boards of directors and supervisors must again consider the risks of investments in fossil fuels and ignoring the climate targets.

It's likely that many of these climate damage cases will fail – they're entering new territory and the accused will resist tooth and nail because they think that they are being unfairly pursued. But only one such case needs to succeed for it to set a precedent.[18]

So in all likelihood, the wave of climate lawsuits will be 'a material driver of the energy transition', concludes the report *The Carbon Boomerang*, which lists all the cases. That's precisely what Roger Cox hopes to achieve. 'You can come up with a thousand supposedly good reasons why we're not doing what we should be doing,' he says in one of our conversations. 'But actually not one of them is valid when you see what's at stake.'

Investors beware

It's not only climate lawyers who are so outspoken. 'If it is wrong to wreck the climate, then it is wrong to profit from that wreckage.' That quote from May Boeve, executive director of the action group 350.org, sums up in a single sentence the starting point of the second strategic campaign now radically changing the world.

This campaign is carried out by the activists of what is known as the 'divestment' movement. Divestment is the term for getting rid of investments because of moral or political objections. The leaders of this campaign appeal to pension funds, cities, states, universities, religious institutions, care providers and museums to cut their financial relationships with the fossil industry.

The divestment activists target the hundred companies responsible for 70 per cent of all emissions that have entered the air since 1988.[19] It was consumers, cities and factories that greedily consumed the fossil energy, that's true. But it was a select group of fossil giants that got the lion's share of the fuels out of the ground – well aware of the risks.

You might say that's all in the past. Since then, every oil company has acknowledged the seriousness of global warming. For Shell it's even 'the most important issue'. If you go to a Shell shareholder meeting, you'll find yourself stumbling over good intentions. The company wants to be a 'force for good' in the world. Exxon is investing in algae as a basis for future fuels. Collectively, the companies give the impression that they are doing everything they can to adjust.

But in the meantime, their investments tell a very different story. The market-listed oil companies are still using the vast majority of their knowledge, money and (lobbying) power to sniff out and drill for new fossil reserves. The fossil giants continue to send ships out to sea to look for *new* reserves, while we have long known that if we burn the known fossil fuel reserves, we can kiss the climate targets goodbye.

In concrete terms, using up the known supplies of oil, gas and coal would cause the emission of more than 3,500 billion tonnes of CO_2.[20] But if we want a reasonable chance at keeping global warming below 2 degrees, we can't emit more than 570 billion tonnes.[21] That's less than a fifth of the current reserves. The rest needs to stay in the ground.

That compelling logic gave rise to the divestment movement. A growing group of activists claims that it's not just because of climate lawsuits that it's imprudent to continue to invest in fossil energy. It's also morally indefensible. It's reprehensible when public institutions and investors continue to facilitate exploitation with their investments in the fossil industry.

The divestment campaign around fossil energy began in 2010. It started on a campus of Swarthmore College, Pennsylvania. A group of students was researching coal extraction in West Virginia and was utterly shocked. Entire mountain peaks were blown up with explosives to get coal out of the ground (unpoetically, in the industry it's called 'mountaintop removal coal mining'). Drinking water and rivers were polluted by the poisonous chemicals released in the process. Something needed to be done.

The students focused on the investment portfolio of their university: Swarthmore invested in coal companies. What if they were to convince the university management to sell those shares and make a public statement about it? In time that could hurt the industry.

The students adopted the divestment model from previous campaigns against the tobacco industry and against apartheid in South Africa. It spread like wildfire.

In spring 2012, campaigns were running on around fifty American campuses. And after a couple of years the movement reached big investors. In September 2014, the heirs of the famous oil magnate John D. Rockefeller decided to publicly withdraw their hundreds of millions from the fossil industry. If the entrepreneur Rockefeller were alive today, he would have invested in sustainable energy, the chair of the Rockefeller Brothers Fund declared.

At the time of the Paris climate summit, in December 2015, the divestment campaign had already launched in forty-three countries.[22] Five hundred investors, cities and religious institutions pledged to make their investment portfolios entirely or partially 'fossil free', which means disposing of investments in the fossil industry.[23] Coal was the first thing to get axed – it's the dirtiest fossil energy source – then oil, then natural gas. Some investors opted to keep their interests in relatively 'clean' oil and gas companies; others broke all financial bonds in one go.

Another three years later, in September 2018, in total almost a thousand institutional investors with a combined capital of $6.24 trillion had agreed to go 'fossil free'.[24]

In March 2019, the Norwegian state investment fund – which has grown rich on profits from oil – announced that it was disposing of its almost $8 billion of investments in oil and gas companies, having previously withdrawn all investments in coal.

Every euro released in this way can be invested in sustainable energy and other climate solutions. It's the beginning of the biggest capital shift in human history.

*

But for every triumph there is at least one defeat. The very first divestment campaign, at Swarthmore, wasn't directly successful: the chair of the board didn't believe that stigmatising fossil fuels would lead to reduced CO_2 emissions. The number of campuses in the US that have made their investments fossil free has since risen to forty-eight. Out of a total of more than 800 universities with investment funds, that's rather meagre.

The world's top three asset managers – BlackRock, State Street and Vanguard – still oversee a combined $300 billion in fossil fuel investments.[25] Pension funds the world over are heavily invested in fossil energy companies, and despite fierce local action and repeated warnings about devaluation – from central banks among others – they are taking their sweet time to scale back those interests.

Does that make divestment campaigns pointless? No. The industry's image certainly suffers under the continuous actions. Take Shell, for example. Loss of confidence is 'the biggest challenge we have at the moment as a company,' CEO Ben van Beurden said in 2017. 'Societal acceptance of the energy system as we have it is just disappearing.'[26]

That's frustrating for Shell, but good for the climate. Because if the image of a company or sector is under pressure, that affects *everything*. Employees would rather opt for a different employer, and governments have fewer qualms about introducing policies that affect the associated company or sector. It creates more political space for action. It's an indirect effect, unmeasurable and not traceable to direct actions, but it's still of great importance.

The divestment movement isn't just about dumping fossil investments. The activists also argue for cutting other connections which give the industry status, or help it progress. If they find out, for instance, that BP sponsors the National Portrait Gallery, the divestment campaigners swing into action. Is Rotterdam's Erasmus University rather chummy with the fossil industry? Action. Is a German insurer underwriting a Polish coal mine? Action.

In part thanks to this activism, more and more people see that what the fossil energy companies do is morally ropy, or even reckless. Commercial bank HSBC recently advised investors to dump fossil energy shares, and warned that anyone who neglected to do so might end up on 'the wrong side of history'.[27]

This is the language of the divestment campaign coming out of the mouths of Wall Street bankers.

And the divestment movement is growing. At the end of 2017, one of the richest cities on earth – New York City – decided to withdraw all investments from the fossil industry. In July 2018, Ireland and Sweden joined the movement, becoming the first countries to withdraw all public funds from oil, gas and coal.

Then central banks caught on. In September 2019, as Christine Lagarde made her pitch to the EU Parliament to head the European Central Bank (ECB), she argued for using the ECB's monumental power to aid the energy transition. 'Any institution has to have climate change risk and protection of the environment at the core of their mission,' she said.[28]

In November 2019, the European Investment Bank (EIB), the EU's financial arm, decided to end funding for fossil fuel energy projects from the end of 2021.[29] Sweden's central bank is already dumping bonds from governments with high emissions, opting for so-called 'green bonds' instead.[30]

So yes, the Great Turn is in motion. The World Bank is no longer financing any new oil or gas projects. A number of the world's biggest insurers are no longer willing to underwrite companies that gain 30 per cent or more of their income from coal. Goldman Sachs has said it will no longer finance new coal mines or power stations, as well as new oil drilling or exploration in the Arctic.[31] And when Saudi Arabia started selling parts of its national oil company Saudi Aramco on the stock market in 2019, international investors were wary of buying in and the public offering had to be scaled back drastically, signalling a loss of trust in the future of oil and gas.[32]

Then corona happened – 'the biggest shock to the global energy system in more than seven decades', according to the International Energy Agency. Demand for oil, gas and coal plunged so deeply that the expectation, even among oil executives, was that it would take years for fossil fuel companies to recover – if they ever would. Perhaps, some speculated, 'peak fossil fuel demand' was already behind us – meaning the sector would only continue to decline from here on out. And with the industry's fragility showcased so bluntly, with renewables becoming ever cheaper and with people continuing to ask for cleaner air, the case for divestment only grows stronger.

Against fossil infrastructure

In addition to the legal battle over the responsibility for climate warming and the divestment campaigns, there is a third strategic campaign for the climate: non-violent direct action to obstruct actual fossil energy projects. Naomi Klein calls it 'Blockadia': the motley collection of worldwide actions to block pipelines, oil drilling and other fossil infrastructure.[33]

This form of action is far from new. Activists chaining themselves to oil rigs or bulldozers are a longstanding cliché of the environmental movement. But in recent years it has once again been demonstrated how effective such resistance can be, and how quickly it can spread.

Blockadia was present in France in 2011, when companies were hoping to start fracking natural gas from shale rock. Activists pointed to the pollution and risks to the drinking water supply that result from fracking. Thanks to them, it was banned.

Blockadia surfaced in 2015, when Shell planned to drill for oil at the North Pole: activists from Greenpeace boarded an oil rig headed there and wouldn't leave. The action grew to such massive proportions that in the end Shell turned back – according to a press release because exploratory drilling in the pole region had found too little oil, but probably mainly out of fear of further reputational damage.

And Blockadia has been doing its best for years to prevent the expansion of existing coal mines and construction of new coal power stations in Bangladesh, Germany, South Africa and Turkey.[34]

The people on the Blockadia front line won't accept the sacrifice of yet more nature reserves and communities at a time when we should be scaling back our consumption of fossil fuels to spare the climate.[35]

*

Far from all these protests are successful in the narrow sense of the word; lots of projects still go ahead, after political and (military) police intervention. In the US in recent years, for example, there has been a large-scale campaign against the construction of the Dakota Access oil pipeline. Native tribes feared that the drinking water from Standing Rock Reservation would be polluted and decided to obstruct the activities on their territory by building a camp on the construction site. The police were called out and used dogs, rubber bullets and bean bag rounds on the peaceful demonstrators.[36]

It worked like pouring oil on the flames. In December 2016, 10,000 people went to Standing Rock. Hollywood celebrities tweeted their fingers off about the protest, and the world press began to write about it.

Initially it seemed to be working. At the end of 2016, the government institution responsible decided to conduct a new environmental survey – effectively suspending the licence.[37] But one of the first things Donald Trump did when he became president in 2017 was to rule that construction should continue.

A failure? Yes and no. Since May 2017, oil has flowed through the Dakota Access Pipeline. But during the protest, seeds were sown and new actions planned. The 27-year-old waitress Alexandria Ocasio-Cortez from the Bronx in New York was one of the thousands of young people who travelled to Standing Rock at the end of 2016 to take part in the protest. Two years later, she became the youngest ever elected Democratic member of Congress.

'I first started considering running for Congress, actually, at Standing Rock in North Dakota,' she said afterwards. 'It was really from that crucible of activism where I saw people putting their lives on the line ... for people they've never met and never known. When I saw that, I knew that I had to do something more.'

Ocasio-Cortez now fights for the Green New Deal from the House of Representatives.[38]

*

It's impossible to gain an overview of all the effects of the three strategic campaigns, but it's clear that they are interconnected and inspire one another. If nature is a web of life, society is a web we've made. We can cut threads in that web and make new connections.

The fact that it's impossible to oversee the precise consequences of our actions doesn't mean they're pointless. It's impossible to know how many animal species and forests precisely are saved as a result of people taking action to protect the environment – but we know for certain that their battle has delivered very real victories.[39] The area of protected nature reserves grew between 1990 and 2018 from 9 per cent to 15 per cent of the total global land surface.[40] Nature needs a great deal more space,[41] and in a third of these protected areas it's still under great pressure due to human activities.[42] Nevertheless, in this area there is cautious talk of progress, thanks to the people who have argued tirelessly for it.

We know that since the Urgenda ruling the Dutch government has continued to be tortuously slow in developing a supplementary climate policy. But things would undoubtedly have been worse if Roger Cox, Koos van den Berg and Marjan Minnesma hadn't won their lawsuit.

In each and every collective action, even the ones that fail, participants exchange ideas, learn how to organise themselves and think of new actions for later.

'Ideas are contagious, emotions are contagious, courage is contagious,' the American essayist Rebecca Solnit writes. 'When we embody those qualities, or their opposites, we convey them to others.'

So it went with the Green New Deal. It was dreamt up in 2017 by the Sunrise Movement, a youth movement which has been fighting for radical climate action in the US for several years now. In 2019, all the Democratic presidential candidates had lined up behind the plan.[43] And then, across the Atlantic, the European Commission launched its own Green Deal to make Europe cleaner, more liveable and more prosperous. When corona hit, the Commission doubled down. 'We should bounce back better from this pandemic', said its President, Ursula von der Leyen, adding that the Green Deal would become 'Europe's motor for recovery.'[44] In the US, the presumed Democratic nominee Joe Biden said he wanted to aid recovery with a 'trillion-dollar infrastructure program' that could 'create good-paying jobs and deal with environmental things.'[45] That sounded very vague indeed, but he might not have said it at all were it not for the momentum created by the Sunrise Movement.

Individual actions can fail, but long term collective efforts are never without a result. It's thanks to the small group of visionaries and activists on the front line that it's now conceivable that we can do what's required in the next thirty years. That's one part of the new story we can now tell each other: we've demonstrated the need, argued the case, enforced the targets and pointed out the alternatives.

The most important question is how we accelerate the turn. How do we build on the momentum that our forerunners have created?

Chapter Nine

THE POWER OF SMALL CHANGES

'I don't like the idea of people always being able to get in touch.'

'I don't see the point of it.'

Two random reactions recorded by a documentary maker in 1998 when he asked passers-by in Amsterdam whether they'd want to be available at all times.

Few could have predicted that twenty years later around 80 per cent of all people in Western Europe would own a smartphone and be online practically 24-7. The age of Wikipedia, overflowing inboxes and social media was imminent. In less than half a human life, just over half of the global population gained access to the web. Change can come fast when we recognise the benefits of something. And when the matter is urgent, as it is now.

Our common objective has been established: halving global greenhouse gas emissions between 2020 and 2030, followed by a linear reduction to zero emissions by 2050. It's only through such a rapid transformation that we can keep well below 2 degrees of warming. And rapid here really means rapid. Three more football World Cups: those of 2022, 2026 and 2030. By then, emissions must have halved worldwide. Children who are born during the 2030 tournament could let their twentieth birthday party coincide with the global festivities that are bound to happen if we meet our target of zero emissions. The truth is, 2050 is just around the corner.

I don't think we've ever faced such a complex task before. Mobilisation for the Second World War was quite a challenge, but at least the enemy back then was an army, an ideology, a Hitler who could be vanquished. The climate isn't something we can simply overcome. Are there any similar challenges? Nuclear disarmament, perhaps – another existential issue. But that's something that a small group of world leaders can, in principle, reach agreements on. It's not happening enough, but it's a possibility. And corona? The damage is incredible but a single vaccine can fix a great deal of it.

If we are to eliminate CO_2 emissions, every single factory will have to be refitted, every petrol car scrapped, and we'll have to start digging up just about every neighbourhood to install new heating facilities. The challenge is truly unprecedented.

But so are the potential gains if we do our best. And we're not just talking about putting a halt to the warming. The basis has been created for an inspirational green narrative, an epic turnaround, a change like no other. Clean energy and clean air for all, a massive economic boost, more peaceful geopolitical relationships, care for one another and for nature – all this we can achieve this century. From here to Timbuktu the Great Turn is underway – the question now is how we can expand and accelerate it.

How can you join in? Which actions are useful and why? Below I'll outline a few concrete steps you can take to help boost the transition to a sustainable economy.

Influencing politics

The first and most crucial step is putting pressure on the government. Most people believe that the state must set targets and make sure they are met.[1] Fair enough. But that will only happen to a sufficient degree if we elect politicians who are fully committed to this.

The next decade will see three more national elections in most Western democracies (not counting any governments collapsing in the meantime). That gives us about three cabinets' worth of time to halve our emissions. So our votes matter tremendously in the immediate future, and the same is true for local and – for those of us still inside the EU – upcoming European elections. The revolution takes place on all political levels.

Why the onus on the government? Only governments can

introduce rules that apply across the board. The state can introduce minimum standards to phase out polluting practices, use taxation to price emissions and provide extensive support for innovation. A rapid green transformation will only come about if we demand that our government pulls out all the stops.

Indispensable measures such as high CO_2 levies for corporations and support for farmers who wish to become more sustainable will only be introduced if politicians can win a majority to do this. While such 'green' politicians can be found in all parties, unfortunately not all parties show the same willingness to act – and to act fast.

In recent years, we've seen that left-leaning and progressive parties are more committed to introducing environmental measures than right-wing and conservative parties. On the whole, the latter tend to prioritise protecting the established order and to adopt a wait-and-see attitude with respect to the climate.

Political parties that deny or downplay global warming are confined to the far right. While still in the minority, they do leave their mark on the climate debate. That's why voting once every four or five years isn't enough – in the meantime, too, the issue has to remain at the forefront of the political agenda. Joining a political party and having a say in its programme and political course is one way of exercising influence.

And then there's the street. Demonstrations have been shown to be effective in expressing support for climate action. As the physical expression of a widely shared concern, they keep the pressure on. The best recent example is the global wave of protest actions in response to the school strike initiated by Greta Thunberg from Sweden.

The then 15-year-old Thunberg began her 'climate strike' on 20 August 2018 – all by herself. She was inspired by students from the US city of Parkland, who refused to go to school because of the shooting that happened there.

Her parents had misgivings about the strike and tried to dissuade her, but Thunberg was so determined that they eventually relented. She spent an entire day sitting in front of the Swedish parliament with her placard and handed out flyers with the text, 'I am doing this because you adults are shitting on my future.'

That outspokenness became her trademark. On that first day she was alone, but from day two other students started joining her. The protest grew and spread via social media to all four corners of the world. By December of that year, more than 20,000 students in towns and villages around the globe had joined her movement.

That same month, Thunberg was invited to speak at the UN climate summit in Katowice in Poland. She was ruthless as she held up a mirror to the political leaders in the auditorium. 'Our biosphere is being sacrificed so that rich people in countries like mine can live in luxury. It is the sufferings of the many which pay for the luxuries of the few.' She continued with a look ahead to the future: 'The year 2078, I will celebrate my 75th birthday. If I have children maybe they will spend that day with me. Maybe they will ask me about you. Maybe they will ask why you didn't do anything while there still was time to act. You say you love your children above all else, and yet you are stealing their future in front of their very eyes.'[2]

Her speech shot through the global media landscape, and the protest grew apace. On 17 January 2019 some 12,500 students took to the streets in Belgium; a week later their number had risen to 35,000.[3] Students everywhere were announcing strikes. Education ministers in countries around the world were saying that climate 'truancy' wasn't allowed. Thunberg responded to one of these reprimands on Twitter: 'Ok. We hear you. And we don't care. Your statement belongs in a museum.'[4] Was she being unreasonable? Or was this finally someone who articulated the urgency of the climate crisis? The Nobel Committee decided the latter. On 14 March 2019 Thunberg was nominated for the Nobel Peace Prize. At the planned global #schoolstrike a day later more than a million students took to the streets, in 125 countries, on all continents (except Antarctica).[5]

In the end, Thunberg didn't win – perhaps she was too polarising a figure for the Nobel Peace Prize after all – but the protest continued to grow. On 20 September 2019, millions of adults joined the striking youngsters, culminating in the biggest climate demonstration ever.[6] And while Thunberg was conquering the world, the Extinction Rebellion movement, with its stated aim of civil disobedience, started making headway from its base in Great Britain. In the autumn of 2019, the 'XR' protesters used peaceful actions to demand attention – or even declare a state of emergency – for the climate in London, Amsterdam, Dublin, New York and other cities.

Imagine just how many climate conversations were prompted by all these protests, and how many people were inspired to take the next step as a result. Perhaps they became a co-plaintiff in a climate case, for example, a donor to an environmental organisation or a volunteer in a local divestment campaign.

Obviously, some people were put off by the fierce rhetoric of Thunberg and other campaigners. Even I was at times. But friction is part and parcel of change too.

Many revolutionary changes in recent centuries came about after people refused to accept the status quo. The abolition of slavery, women's suffrage and the formal end to child labour were not instigated by politicians, but only happened because social movements forced them to take the necessary action.

Thunberg's greatest inspiration is Rosa Parks,[7] the African-American woman who refused to give up her seat to a white passenger in 1955. It kickstarted the Montgomery bus boycott, a major step on the road to formally ending racial segregation in the US.

What activists and visionaries do and say is seen as unreasonable at first – until it becomes acceptable and sensible. Measures that appear to be radical at first are then broadly welcomed. Only then do politicians turn them into policy.[8]

Thunberg's protest shows how rapidly a movement can grow, but you don't have to be a Thunberg to put pressure on politicians. Nor do you have to be a million strong to exert influence. Politicians also react to smaller local protests, to citizens who ask critical questions of MPs or local councillors. This is what led to the suspension of fracking in the Netherlands, for example. (In the UK, the ban on the extraction of shale gas only came after both protests *and* earthquakes.)[9] Democracy is a constant dialogue, and there are plenty of moments when you can make your voice heard.

How, I hear you ask? Joining the conversation at a local level, casting your vote for rapid greening and taking to the streets are the first three powerful things that any citizen (aged 18 and over) can do.

Will that be enough? I'm afraid not. Too often, climate policy remains a divisive issue, with politicians exaggerating their differences of opinion for electoral gain. For democracies to implement an effective climate policy that can count on sufficient support, they need to be renewed.

Various countries are now experimenting with new democratic methods such as citizens' assemblies. These are made up of people randomly selected from all levels of society (thus guaranteeing diversity). Such panels, of a hundred or more people, are given the opportunity to sink their teeth into a particular issue, hear experts and then table concrete proposals. In the ideal situation, rules are in place stipulating that parliament and government have to respond to the panels' proposals. And if they don't adopt them, they'll have to give a written explanation, after which the dialogue continues.

Following the *gilets jaunes* protests in France, Macron asked such a citizens' assembly to design new climate policies to radically lower emissions 'in a spirit of social justice'. Macron has promised to submit the assembly's proposals 'without a filter' either to 'a referendum, a vote in Parliament or direct implementation'.[10] Of course, he's hoping that when the resulting policy is eventually implemented no protestors take to the streets.

A similar citizens' assembly has begun work in the UK.[11] The corona-crisis has created considerable delays to these processes – large scale gatherings were not exactly prudent. But there will be a post-corona and the conflicts that arise when dealing with climate change still need to be dealt with. Involving citizens in climate measures is the best way forward.

Why might this work? Because it gives citizens shared responsibility for the policy. They can propose long-term solutions and, free from the distraction of the next elections, make the kinds of painful decisions that elected politicians shy away from.

If such panels were to be introduced not just at a national, but also at a local level, they might offer an opportunity for a proper conversation about our society's new green design. While discussing the best possible climate policy, we might be able to rediscover what it's like to work collectively on something, to have a shared goal and not to fall at the first hurdle.

And all this under immense pressure: we have ten years before we're supposed to have halved our emissions. We have no time to lose.

Changing your environment

Politics isn't the only arena where action is both possible and useful. A lot of change starts at a personal level. Look at Marjan Minnesma, the director of Urgenda. She has shown repeatedly what's achievable. And not just by taking legal action to remind the Dutch state of its obligations.

As early as 2009, Minnesma brought fifty large companies together into a cooperative to import electric cars – something that was quite rare up until that point. In 2010, she travelled to China to buy 50,000 solar panels – the first collective purchase of solar panels in the Netherlands, and big enough to crack open the spluttering domestic market. In 2012, she was the co-founder of ThuisBaas, now an independent company that makes homes energy-neutral with the help of solar panels and solar boilers, heat pumps and induction hobs – for the amount of money you'd normally spend on energy in fifteen years, your energy bill will be reduced to zero. In recent years, Minnesma and Urgenda have come up with idea after idea to limit the emissions from agriculture, buildings and industry. And she doesn't just talk the talk: she walks the walk, and shows that it's actually possible.

Positive energy is infectious, she told me. She lives in an old farmhouse in the northwest of the Netherlands, where she hosts New Year's drinks for members of the local neighbourhood association. Although she gives hundreds of lectures around the country every year, Minnesma says that she's never been particularly 'preachy' in her immediate vicinity. But the fact that her own roof is covered in some seventy solar panels has been an inspiration in and of itself. In 2009, she was the first to have such panels, and now at least ten of her neighbours have them too. After making his entire house energy-neutral, one of Minnesma's neighbours approached her to discuss how the rest of the neighbourhood might be weaned off gas.

Of course, not everyone is in a position to implement such measures – a third-floor rental flat can only become sustainable if the government makes it possible or mandatory, something that will have to be done on a large scale in the decades to come. But those who do have the money and the means to produce green energy or to insulate their home will inevitably inspire others. Numerous studies have shown that people are more likely to buy solar panels when their neighbours already have them. The phenomenon is known as 'solar envy', and electric cars inspire a similar response.[12]

Local change triggers more local change. Everywhere you look, pioneers such as Minnesma have blazed a trail. In recent years, countless local initiatives have got off the ground, from energy cooperatives to urban farming projects. Sports clubs and schools are covering their roofs with solar panels, and more and more organisations and companies are aiming to become fully sustainable.

So joining the revolution often means linking up with a local initiative that's already underway. You don't have to live like a monk or turn your entire life upside down, nor do you have to wait for the government to introduce the necessary policies. Only you can find out where you can make the biggest difference. Just remember: if you join a neighbourhood initiative, you don't have to start from scratch. As a team you can make a difference straight away. You're helping to write the epic story of all those neighbours, streets, villages, cities and countries that are leading the way in the Great Turn.

In a crisis you change your behaviour

As well as putting pressure on politicians and furthering local initiatives, you can have a direct influence by changing your consumption patterns. People are often a bit condescending about this, as if such small changes don't help. Obviously, some are totally pointless. We all know that most fish don't benefit from a Dolphin Safe label.

But some changes do matter, because they lead to collective change. Let me give you three examples.

1. *Eating more plants and less meat.* Right now, the world is home to some 1 billion pigs, 22 billion chickens, 1.2 billion sheep, 1 billion goats and 1.5 billion cattle.[13] Those numbers are insane, and the raising of all those animals has led to huge environmental problems. Between half and three quarters (!) of all agricultural land is dedicated to the production of cattle feed and grazing. That excessive land use makes the livestock sector a major driving force behind deforestation and the decline in animal diversity. Some 15 per cent of all greenhouse gases that we emit can be directly traced back to animal husbandry – especially that of cows belching and farting methane.[14] Indirectly, meat consumption is one of the key drivers of the depletion of fresh-water stocks, as a lot of water goes into the production of cattle feed and the raising of the animals themselves. And, of course, keeping animals makes us vulnerable to the diseases that develop in them and then cross to humans – not to mention the strains of antibiotic resistant bacteria evolving in livestock.

Does that mean that the entire livestock population should go? No. There's a place for farmed animals in a sustainable food system – they can eat food scraps and graze on areas of land that won't sustain much else besides grass. But in a liveable future we can't remain the carnivores we are now. That's simply impossible: the costs are too high. Not just the costs to the environment, but those to our health as well. Excessive consumption of red meat – the collective name for the meat of cows, pigs, sheep and goats – is a contributing factor in cancer, strokes and heart failure.[15] A significant section of the global population consumes more red meat than necessary; Southern Asia and sub-Saharan Africa are the only areas where people are below the 'reference intake'.[16]

Compare that to a plant-rich diet with generous quantities of fruit, vegetables, whole grains, nuts and pulses.[17] Switching to a vegetarian diet with limited amounts of milk, cheese and eggs can lead to a 45 per cent reduction in global greenhouse gas emissions of the food sector by 2050. Switching to a fully vegan diet would make it a 55 per cent drop.[18] In both cases, agriculture would consume considerably less water and land.[19]

So if you're still eating meat and you're looking for a way to reduce your impact on the environment: here it is, the most effective step you can take this very instant. And you're not alone if you decide to do so. Around the world, more and more people are switching to a diet with plenty of plants.

In the coming decades, plant-based diets should become the norm, and one reason for believing they will is that it's becoming easier and tastier to cut out meat. Because a growing number of people are changing the way they eat, the market for meat substitutes is rapidly expanding.[20] More money is being invested in the further development of these products, making them both more flavoursome and cheaper. In the decades to come, cultured meat will probably become an affordable alternative.

And the benefits aren't confined to the environment. A global switch to a plant-based diet might translate to a worldwide drop in annual mortality rates of 6 to 10 per cent. The healthcare savings alone are estimated to be $1 trillion a year by 2050.[21]

So there really isn't anything 'small' about your dietary choices. What you put on your plate connects you to the rest of the world. Every meal is a vote for or against a liveable earth, for or against the ongoing development of meat substitutes, for or against a world in which all the livestock added up weigh fourteen times more than all the wild animals together.[22]

It's like we're having elections three times a day.

2. *Taking your money out of fossil fuels.* Did you know that you could well be investing in the extraction of more oil, gas and coal?

The number of people who do this directly and knowingly – as shareholders of a fossil fuel company – is small (if you own such shares, you'd better sell them soon and reinvest the proceeds in a sustainable fund).

But there's a surprisingly big chance that your energy supplier, bank or pension provider may be investing your money in fossil fuel. It's likely that every time you pay your energy bill your household supports a company that continues to generate power with coal and gas.

Government policy plays a decisive role in cleaning up the energy supply, but that's not to say that you're powerless as a consumer; you can still send a signal. Luckily it has become easier in recent years to make sure that your money flows in the right direction. With the help of rankings such as those of consumer organisation Ethical Consumer, you can check how responsible your energy supplier, bank and pension fund really are.

It's easy to switch to an energy provider that only produces electricity with wind turbines and solar panels. If you choose an energy provider or cooperative that invests locally in green power, every payment contributes to the green transformation.

There are banks that promise to invest every pound you save in sustainability – you can switch via currentaccountswitch.co.uk.

And – if you're paying into a pension – you can find out whether your old-age provision is already being fully invested in green funds. ShareAction and Fossil Free UK can give you the necessary information. If you're still building up a dirty pension, you can join these organisations to change this.

Small businesses can sign up to a clean auto-enrolment pension scheme launched by Ethical Consumer and ethical financial advisory group Castlefield. Having said that, it's always sensible to seek advice from an adviser before you make any radical decisions about your finances. Set aside an hour or so in your diary to get started. With every pound that you commit to sustainability, you accelerate the capital shift that has already been unleashed by the divestment campaign.

3. *Start greening your travel.* Some 15 per cent of all greenhouse gas emissions are caused by the transport sector, specifically by the burning of oil to power container ships, aircraft, lorries, cars and motorbikes. Within this category, urban transportation is the worst culprit. With no fewer than 54 per cent of the global population living in cities, many of us are in a position to make a daily impact on this source of emissions.

Our transport choices have consequences not only for the climate, but also for the air quality. According to the World Health Organization, nine out of ten people around the globe breathe polluted air every day.[23] Exhaust fumes are a major contributor to this pollution, but there are things we can do.

Electric cars have become a progressively more affordable alternative to petrol cars. Public transport and bikes, electric or not, are even better. These options aren't yet universally available, but those of you who do have access can make a difference by using them. Every time we opt for those alternatives, we avoid emissions. And the more people leave their cars at home, the more space is freed up in the city for pedestrians, homes, businesses and greenery.

A smaller, but growing source of emissions that we also have a direct influence on is aviation. At just over 2 per cent, it only represents a limited share of overall global CO_2 emissions. But the sector is growing everywhere, and is expected to account for a fifth of all emissions by 2050. That growth is incompatible with meeting climate targets, just as it flies in the face of a climate-friendly lifestyle.

Did you decide to give up meat for the sake of the climate? One return flight from London to New York is as bad for the climate as consuming almost 1,000 Big Macs.

Have you exchanged your lightbulbs at home for environmentally friendly LEDs? The CO_2 you will save over five years is cancelled out by one medium-haul flight, from, say, Berlin to Lisbon.[24]

I'm not saying this to suggest that vegetarianism and energy saving are pointless – on the contrary, these are and will continue to be effective steps. But we must acknowledge the uncomfortable truth that if we continue to fly, we undo the progress in many other sectors. After all, it takes a lot of energy to lift a metal bucket full of people and suitcases into the air.

Go figure: a Boeing 747 needs more than 190 tonnes of kerosene for a long-haul flight. With 410 people on board, that's four bathtubs per passenger. If the aircraft burns nearly all of that fuel during a long flight, it emits 530 tonnes of CO_2. That's nearly 2.5 times the weight of the plane, all simply deposited in the atmosphere.[25] Those trails produced by aeroplanes? That's frozen condensation. For each molecule of water, another molecule of CO_2 is left behind.[26] But *that* we can't see.

We don't all fly the same amount. In the UK, for instance, 'the 10% most frequent flyers took more than half of all international flights in 2018,' one survey found. 'Just 1% of English residents are responsible for nearly a fifth of all flights abroad.'[27] In the US, '12 per cent of Americans who make more than six round trips by air a year, are responsible for two thirds of aviation emissions', according to a recent analysis.[28]

If you belong to this group, you could consider booking a train, choosing a different holiday destination or conducting your next meeting over videocall. Is that long flight for work or a family visit really unavoidable? In that case you could compensate for your emissions by supporting certified reforestation projects – there are various guides online and providers including goldstandard.org and standfortrees.org offer reliable schemes. Better still: compensate tenfold to increase the chance that the stretch of wood that you're protecting or planting will survive long enough to ensure net compensation for your actual emissions. In this way, you pay the real cost of the climate damage you're causing.

The power of small changes

Our decision-making is not as elegant as that of the bees, but when all is said and done, we also carry out quite a few future dances. These take the form of the stories we tell one another about where we're headed, our choices in the voting booth and in the supermarket, and the organisations and initiatives we support.

There are so many ways of doing *more*. If the world's richest 10 per cent were to adopt the consumption pattern of the *average* European, global emissions would drop by a third.[29] If you were to shop more smartly and waste less food as a result, you'd contribute to reducing today's mind-boggling food waste – in total, nearly one *third* of all food is wasted.[30] If more people were to adopt a different lifestyle, there'd be a lot less need to use 'scrubbing technology'

– such as biofuels in combination with the underground storage of CO_2 – later this century.[31]

All of these behavioural changes are beneficial for the climate. Obviously not if you were the only one to follow through: based on figures alone, it doesn't make any difference whether or not you prevent CO_2 emissions, since your share is negligible among a total of 7.8 billion people. I know from experience that you can feel very lonely with your good intentions and your shrinking footprint.

But in reality you're never alone. With every step you take you're part of a growing group that realises that we need action and behavioural change – and that we need them fast. When you change your behaviour, you become part of a transformation that millions of others are already working towards. And who knows, maybe even billions soon. You're helping to realise the green narrative and a crucial cultural shift. You don't 'lose out' by making changes, because your actions lead to progress: greater well-being, cleaner air, a better climate. The main purpose of sustainable behaviour is not a pure, green life at an individual level. Nor is it a matter of being 'better' than others or totally eradicating hypocrisy. At the end of the day people are fallible, and besides, within the current system it's impossible to avoid any negative impact on nature. That's why a change in consumption is never enough – at best, it's a minimum. The emphasis should always be on individual choices translating into collective ones. And that's happening.

Every sustainable choice is a signal that can lead to political change. It will be easier for the government to introduce a flight tax, for instance, when more and more citizens are prepared to align their travel patterns with the gravity of global warming. More people using international high-speed trains more often will strengthen the business case for expanding the network. Put differently: individual choices always have collective consequences. It's then up to the government to make sure that sustainability standards become the new normal for us all. But until that happens, none of us has to be idle: every single tonne of CO_2 emissions that's avoided matters.

The small group of people who bought a mobile phone in the

late 1990s made the subsequent development of the smartphone possible. The handful of individuals who bought the first exorbitantly priced Teslas paved the way for cheaper (but still expensive) models. And thanks to the emerging competition, the entire car industry is now caught up in a race to produce the cheapest electric vehicle.

You can call it the power of small changes. That power is not confined to the market; it's felt in politics too: it only takes a small percentage of voters to change their mind and another party will emerge as the biggest.

That's not to say that it's going to be easy. At present we're not observing the first rule of holes: if you're in one, stop digging. In 2018, for instance, car emissions increased worldwide, despite the rise in the number of electric vehicles on the road. The reason? People also bought more SUVs.[32]

It's these kinds of facts that, on bad days, make me feel hopeless. But then I remind myself that we don't know how this story ends. The behaviour of CO_2 particles is predictable, that of people is not. For evidence just look, one more time, at the global response to the corona-outbreak. At the outset, who would've thought that in response to this deadly virus we would be showcasing solidarity on a massive scale? We did just that by staying home, radically changing our behavior so that other people we don't even know wouldn't get sick. Will it change us? Perhaps. 'A new awareness of how each of us belongs to the whole and depends on it may strengthen the case for meaningful climate action, as we learn that sudden and profound change is possible after all,' writes Rebecca Solnit.[33] I hope it does.

In the end, nobody knows whether or not we'll do enough to prevent a hothouse earth. Or whether we'll be able to reverse the decline in species diversity. But that we have it in us to turn the tide is certain. Man is capable of change – over the course of history our species has never stopped evolving. And we keep on inspiring one another.

This is perhaps most apparent in our relationship with food. Roger Cox, the lawyer I discussed at length in the previous chapter, gave me a good example. When his elder daughter was six, she became a vegetarian. The rest of the family was still eating meat, but Cox's mother followed suit and became a vegetarian herself two years later. His younger daughter then caught the veggie bug and at some point even refused to carry a shopping bag that contained meat. Now the family is eating vegan and enjoying tastier meals than ever, Cox explains.

This is not uncommon. One day, aged twelve, Greta Thunberg decided to stop flying. When her family asked why, Greta explained calmly and cogently what her motivations were, pointing to the climate science findings which had made her see that this was a serious issue. She showed that trains could get her places too. Some time later her mother came round to her way of thinking, with major ramifications for her career as an opera singer. And then her father and sister did the same.

This is how Thunberg puts it: 'When you are in a crisis, you change your behaviour.'[34]

Or you don't change your behaviour, of course. But at the end of the day, standing on the sidelines is also a choice. That, too, has an impact.

We often gloss over the consequences of inaction. We talk about the climate as if we're in the stands and the players on the pitch are making a mess of things – look at them floundering, they can't get their act together. But in this match we're all players – and that includes the journalists. That's why I made a point of not standing on the sidelines in this book, instead adopting principled positions and focusing on solutions – because you simply can't remain neutral. We're all on the pitch and that pitch is the earth, and the people who are supposedly 'on the sidelines' have influence too. They maintain the status quo by continuing to act as if all is normal.

But there is no more normal; it's receding in our rear-view mirror. The planet we're living on is a bleaker variant of the one we grew up on. We can and will adapt. We can avoid worse. We're writing our own future. It's what we did during the Agricultural Revolution, it's what we did during the Industrial Revolution, and it's what we're doing again now. There's no blueprint; all we have is the knowledge that there's only one earth, that we all depend on one another, and that as a result everything depends on what we think, dream, say, dare and do.[35]

You don't have to *feel* part of the story-writing collective to *be* a part of it. As the South African poet Antjie Krog put it, 'One cannot choose not to have an umbilical cord.'[36]

You can choose to take part in your own way.

APPENDIX

12 CLIMATE-RELATED MISUNDERSTANDINGS

In conversations and debates about global warming, fiercely held opinions are flung back and forth. That's good – exchange of ideas is the fuel of progress. But too often it's no more than the reiteration of old misunderstandings or untruths. So here are the facts on twelve climate-related misunderstandings.

1. The climate is always changing, so the current warming is no big deal.

It's true that the climate is always changing, but that's no reason to trivialise the current changes. It's like setting your car on fire and then saying the engine temperature has changed before.

Over the past 12,000 years, the climate has been relatively stable, warm and wet. We're now leaving that stable climate niche and risking a hothouse earth. That's dangerous, because the heat we're now charging towards is at a level we've never experienced before.

Never before has the climate changed as quickly and radically, and now there are 7.8 billion people living on the earth. All those people will be burdened with higher temperatures, rising sea levels and more extreme weather.

2. We don't know if it's our greenhouse gases causing global warming.

On the basis of countless observations, climate scientists have established this with great certainty: human emissions of greenhouse gases are the dominant cause of current global warming. That's because greenhouse gases such as CO_2, methane and nitrous oxide retain warmth in the atmosphere.

Measurements show that the average temperatures at night are rising faster than daytime temperatures.[1] That's indicative of the greenhouse effect: the blanket of gases around the earth prevents heat from escaping into space. So at night things cool down less than they did before we began to emit greenhouse gases on a massive scale.

Other explanations for the current warming climate have been refuted. For instance, it cannot be caused by extra solar radiation. Although there is certainly some minimal variation in the quantity of solar radiation reaching the earth, that variation is far too small to cause the current climate change. Moreover, that variation works in both directions, while the earth becomes warmer year after year.[2]

Only the increase in greenhouse gases can explain all the changes we're now experiencing.

3. The consequences of climate change are positive. Plants love CO_2 and there will be new arable regions around the North Pole.

The second sentence is right, the first isn't.

Plants and trees do use CO_2 to grow, and melting ice, for instance in Greenland, may free up land for new activities – from extraction of raw materials to agriculture. But such relatively small – and often dubious – benefits don't cancel out the gigantic drawbacks of global warming.

Many negative consequences of global warming are already manifesting themselves: from the growth of hypoxic zones at sea and increased damage from storms to the increase in potentially deadly heatwaves in summer.

Even the food supply is under serious threat from global

warming, cities with millions of inhabitants lie in the danger zone due to rising sea levels, and low-lying islands will disappear. In the coming decades, millions of people will be displaced by the consequences of global warming.

If the earth continues to heat up, the vast majority of humankind is worse off, and the risks of extreme climate change are uncomfortably big.

4. It doesn't matter what a relatively small country does. There's no point unless China and India do something.

Imagine the atmosphere as a bathtub. All over the earth there are taps on, big and small, filling the bath. How do we ensure it doesn't overflow? By turning off all the taps.

It doesn't matter how much a specific country contributes in emissions: every tonne of CO_2 not emitted is a step in the right direction. And in order to keep global warming under 2 degrees Celsius, in the next thirty years all the taps need turning off.

Of course it's very important that the big emitters turn their taps off faster. Fortunately, that's the plan. Of all investments in sustainable energy from 2010 onwards, a good 31 per cent came from China. The US stalled at 14 per cent of the total, while all the European countries together came to no more than 28 per cent.[3]

India, which still relies heavily on coal, also wants to become a leader in the revolution of solar and wind energy. In 2018, solar panels supplied more than half the new electricity-production capacity in the country.

Leaders such as Germany, China and India see strategic advantages in a completely sustainable energy supply. On the other hand, the countries that are currently reluctant to open their wallets will be dependent on technology developed elsewhere for their future energy supply. Those who hesitate now will miss out on the jobs and prosperity that come with a green economy.

5. We're lost: politicians who don't want to do anything about climate change are winning.

That's how it seems sometimes, but appearances can be deceptive. Both within and outside politics, the group of people who *are* working to keep the world inhabitable is many times larger than the group completely ignoring the problem or unwilling to act on it.

In June 2019, a survey carried out among 30,000 people in twenty-eight countries showed that a majority of 80 to 90 per cent believe that the climate is changing and humanity is responsible, or partly responsible.[4] About seven in ten people believe 'climate change is a major threat to their country'.[5] Over 90 per cent of Europeans polled in April 2019 think climate change is a serious problem; almost eight in ten think it is a very serious problem.[6]

In 2017, more than 80 per cent of respondents to a questionnaire representative of the entire world population said that they considered it important 'to create a world fully powered by renewable energy'. Solar energy was their top choice.[7]

Major changes always invoke uncertainty, and fear, and the tendency for people to dig their heels in. But such reactions don't in any way detract from the benefits of the course already begun towards a liveable climate, clean air and healthy nature.

6. Spending money on climate policy is a waste: what if global warming turns out not to be such a problem after all?

The better question is: what if it turns out worse than expected? There's no back-up planet.

But imagine that global warming does turn out not to be such a problem (very improbable) and that we adapt very well to the consequences (unpredictable).

There's still the mass extinction of animal species that has been set in motion by our activities, with enormous risks for the functioning of ecosystems on which human communities are dependent for their food and drinking water. Then there's the plastic and the dead zones in the oceans. There are hundreds of reasons for rapidly moving towards sustainability.

And what's the worst that can happen? At worst we will achieve a sustainable way of living, on the only planet we have, slightly earlier than was strictly necessary. In the end, that's in everyone's interest.

7. Climate policy is unaffordable. There are already people who can't pay their energy bills.

The second sentence is true, the first is not.

The costs of sustainable technologies are constantly falling, and the technology is progressing in leaps and bounds. It's therefore ever cheaper to tackle emissions. The net impact of climate policy is primarily very positive: new jobs, new economic activities, increased prosperity.

If we don't invest in prevention now, later on we'll be forced to spend much larger sums on measures to adapt to the consequences of global warming. Floods are expensive. Droughts are expensive. Climate damage is already costing the world economy tens of billions a year, and the bill will only go up as emissions rise.[8] One analysis after another shows that tackling climate change is *much* cheaper than letting it happen.

In the end, the choice as to whose shoulders bear the heaviest burden is a political one. In many countries, sustainable energy is subsidised by a surcharge on energy bills, which hits poor households relatively hard. Anyone who is against that can vote for a party that intends on spreading the burden more fairly.

8. What climate activists want is too radical and therefore unrealistic.

People who worry about the climate, and who take to the streets to protest, generally demand nothing more than the action prescribed by climate science to prevent further climate disruption.

All the experts agree that we have the technological means to switch quickly. The only thing that's radical or unrealistic is to continue pretending that climate change does not exist.

9. We'll never achieve sustainability, because the world economy just keeps on growing at the cost of the earth and big companies only ever choose profit.

Economic growth has so far come at an enormous ecological cost. But that needn't remain the case.

Sectors such as education can already grow just fine without being an additional burden to the environment. Once we've made other sectors of the economy sustainable, growth in itself will no longer be a cause for concern.

Lots of companies are already upscaling their climate ambitions and see growth opportunities in a green economy. In every sector, plans are afoot, and those left behind are feeling the pressure. Not enough is happening yet, but a green course is now visible *and* attractive.

A complete change overnight is impossible, but gradual change over a period of thirty years? That's not only very possible, it's also happened plenty of times before.

10. My contribution is a drop in the ocean.

Every contribution is worthwhile, because the consequences of all our actions are cumulative. People constantly inspire one another: we follow each other's example, we influence each other. All over the world, people have heard the call to action. The forerunners worldwide are working together on a sustainable society. If you decide to join in, you join this powerful movement.

11. People who act sustainably are hypocritical: they're still polluting the environment.

Anyone who makes the occasional sustainable choice is familiar with this type of criticism: 'You're a vegetarian, but you still wear leather shoes.' Critics clearly have a problem with other people not being perfect. But is that a personal failing? Is perfection the goal?

In today's society, it's more or less impossible to live an ecologically 'pure' life. But that doesn't make it right to attack the people who take steps in the right direction. Those trying to make their actions more sustainable are doing it not to feel superior, but to accelerate the transition to a different system. The goal is collective change – and that has to start somewhere. In fact, it's already begun. Those who join in strengthen the current.

12. There are too few of us to make a difference.

There are more than enough of us, because it only takes a small group to influence the course of a whole society.[9] That's been shown plenty of times, for instance in the abolition of slavery and the introduction of voting rights for women.

Anyone working on this shift is part of a cross-border movement to keep the earth habitable. The benefits of a green path are clear, and the urgency is great.

AFTERWORD AND FURTHER READING

Most of this book came into being on *De Correspondent*, a member-based online platform that focuses on independent, in-depth and ad-free journalism. Some passages already have been published there, often in a different form.

The future sketches at the beginning of chapters 5 and 6 were my own invention, but they're partially supported by real research and current developments. For instance, the Amazon region is drying up very rapidly,[1] and fear of the dykes overflowing the Netherlands is a genuine concern.[2] The predicted temperature rises in the two scenarios in the year 2050 are based on the 2014 IPCC report.[3] The term 'Great Turn' was coined by writer Hanna Bervoets.[4]

In recent years, I've been particularly inspired by a number of writers. The biggest source of inspiration for my thinking on change is the American writer Rebecca Solnit. I've learnt a great deal from her columns in *Harper's Magazine* and the *Guardian*. Read her book *Hope in the Dark* (2004) if you think things look hopelessly bad for the world. Read *A Paradise Built in Hell* (2009) if you want to know how people behave in times of disaster.

Bill McKibben is one of the first journalists to have written about the climate. His book *The End of Nature*, published in 1989, remains spot-on. His excellent articles, combined with his activist role in the climate organisation 350.org, have made him a role model for me: McKibben shows that involvement and journalistic curiosity can be combined.

Elizabeth Kolbert, who works for the *New Yorker*, wrote one of the best books on climate change, *Field Notes from a Catastrophe* (2006), followed by an important book on the mass extinction of animal species, *The Sixth Extinction* (2014).

This Changes Everything (2014) by Naomi Klein is one of the most influential books on the climate written in recent years. I read it as soon as it came out, and it made a big impression on me. Klein's lecture 'Let Them Drown' (2014) and her books *No is Not Enough* (2017) and *On Fire: The Burning Case for a Green New Deal* (2019) are also worth reading.

If you want to know more about the consequences of rising sea levels, read *The Water Will Come* (2017) by Jeff Goodell. If you're interested in how the first farmers began to change the climate, *Plows, Plagues and Petroleum* (2005) by climatologist William Ruddiman is indispensable.

Roy Scranton wrote *Learning to Die in the Anthropocene* (2015), a great book for anyone who thinks the climate has gone to pot. He describes very clearly what a bind our societies are in, and was an important source for the first future scenario. I also first read about decision-making among bees in Scranton's work.

David Wallace-Wells, who published the article 'The Uninhabitable Earth' in *New York Magazine* in 2017, and in 2019 brought out a book of the same name, sharpened my thoughts on the seriousness of the climate crisis.

Bruno Latour was also an important source of inspiration. The French philosopher first wrote the provocative *Facing Gaia* (2017) and later *Down to Earth* (2018). In the latter book he uses his philosophical toolkit to find a way forward. Witty and enriching.

Two other fascinating thinkers who have written about nature and humans are Timothy Morton and Donna J. Haraway. If you like philosophy, I'd particularly recommend *Being Ecological* (2018) and *Staying with the Trouble* (2016) respectively.

I view David Roberts as one of the best climate journalists of the moment: he writes one sharp article after another for *Vox* about global warming and what measures we can take to combat it. I have learned a great deal from Roberts in recent years, and from George Monbiot, who writes for the *Guardian*, and Jeremy Leggett, who is tirelessly mapping out the energy transition and also wrote *The Winning of the Carbon War* (2018).

While writing this book I repeatedly reached for *The Human Planet: How We Created the Anthropocene* (2018) by Simon L. Lewis and Mark A. Maslin. A must-read if you want to understand how humans have managed to herald a new geological era.

Origins (2019) by Lewis Dartnell tells the same story, but places more emphasis on how the earth has shaped *us* instead of the other way around. That book, too, helped me hone my thoughts.

The book *Four Futures: Life After Capitalism* (2016) by Peter Frase was a model to me in writing my own future scenarios.

Andrea Wulf's book about Alexander von Humboldt, *The Invention of Nature* (2015), is a must-read if you want to know more about the 'web of life', and about the fascinating von Humboldt.

If you'd like to find out more about the countless options we have to tackle emissions, read *Drawdown: The Most Comprehensive Plan Ever Proposed to Reverse Global Warming* (2017). The book makes good on its title, and provided me with a useful reference point in assessing different measures against global warming.

*

For the foreseeable future, I'll continue my search for new stories and solutions at *De Correspondent*. I'll also continue to report on what's going wrong and keep an open mind throughout. Call it constructive investigative journalism: curious about problems as well as solutions. English speakers can join *The Correspondent* to support the expansion of collaborative, constructive, ad-free journalism. Visit thecorrespondent.com.

NOTES

Preface

1. Philippus Wester et al., *The Hindu Kush Himalaya Assessment*, Springer International Publishing (2019). For a summary see: Chelsea Harvey, 'World's "Third Pole" Is Melting Away', *Scientific American* via *E&E News* (4 February 2019).
2. The study appears here: Tapio Schneider et al., 'Possible climate transitions from breakup of stratocumulus decks under greenhouse warming', *Nature Geoscience*, Vol. 12 (2019), pp. 163–7. For a summary see: Zeke Hausfather, 'Extreme CO_2 levels could trigger clouds "tipping point" and 8C of global warming', *Carbon Brief* (25 February 2019).
3. Natalie Wolchover, 'A World Without Clouds', *Quanta Magazine* (25 February 2019).
4. The name Tom is made up to preserve his privacy.
5. I derive this insight from the Italian philosopher Giorgio Agamben. To him, 'despair is a moment of possible change, and therefore a reason for hope'. Source: Joe van der Meulen, 'Profiel van filosoof Giorgio Agamben: Gids door de hel' ('Profile of philosopher Giorgio Agamben: A guide through hell'), *De Groene Amsterdammer* (2015), No. 27–8.
6. In her essays, Rebecca Solnit repeatedly makes this point about hope. See for example the introduction of: *Hope in the Dark: Untold Histories, Wild Possibilities*, Haymarket Books (2015), p. xvi. 'A victory [in a social struggle] doesn't mean that everything is now going to be nice forever.'

1. How We Got Into Trouble

1. The figure 38 billion is an estimate for 'reference man'. The figure varies from one individual to another. Source: Ron Sender et al., 'Revised Estimates for the Number of Human and Bacteria Cells in the Body', *PLOS Biology*, Vol. 14, Issue 8 (2016).
2. For a good review of the research into declining insect numbers, see

Brooke Jarvis, 'The Insect Apocalypse Is Here: What does it mean for the rest of life on Earth?', *The New York Times Magazine* (27 November 2018).

3. A study in Great Britain found that the press gives eight times more coverage to the climate than it does to biodiversity. Pierre Legagneux et al., 'Our House Is Burning: Discrepancy in Climate Change vs. Biodiversity Coverage in the Media as Compared to Scientific Literature', *Frontiers in Ecology and Evolution*, Vol. 5 (2018).
4. Ewen Callaway, 'Oldest Homo sapiens fossil claim rewrites our species' history', *Nature* (7 June 2017).
5. Yuval Noah Harari, *Sapiens: A Brief History of Humankind*, Harvill Secker (2014), p. 73.
6. For an enlightening discussion of this 'climate pump', see the first chapter of Lewis Dartnell, *Origins: How The Earth Made Us*, Penguin (2019).
7. I have borrowed this image from *Sapiens*, p. 122.
8. Prior to the current warm epoch, the Holocene, *Homo sapiens* lived through another warm period: the Eemian interglacial (from 130,000 to 115,000 years ago). However, this was before the advent of modern man, between 80,000 and 50,000 years ago. This modern human is anatomically very similar to today's man and has greater creative powers than the earlier *Homo sapiens*. See: Simon L. Lewis and Mark A. Maslin, *The Human Planet: How We Created the Anthropocene*, Pelican (2018), Chapter 3.
9. This temperature change can be found in IPCC, *Climate Change 2013: The Physical Science Basis. Contribution of Working Group I to the Fifth Assessment Report of the Intergovernmental Panel on Climate Change*, Cambridge University Press (2013), Chapter 5, and p. 404. Hereafter this report will be referred to as 'AR5'.
10. It's thought that humans crossed over to America some 15,000 years ago, although the estimates vary. See: J. R. McNeill, 'Global Environmental History: The First 150,000 Years', *A Companion to Global Environmental History*, Blackwell Publishing (2012), p. 5. And Mikkel W. Pedersen et al., 'Postglacial viability and colonization in North America's icefree corridor', *Nature*, Vol. 537 (2016), pp. 45–9.
11. *The Human Planet*, p. 101.
12. J. R. Petit et al., 'Climate and atmospheric history of the past 420,000 years from the Vostok ice core, Antarctica', *Nature*, Vol. 399 (1999), pp. 429–36.
13. Ainit Snir et al., 'The Origin of Cultivation and ProtoWeeds, Long Before Neolithic Farming', *PLOS ONE*, Vol. 10, Issue 7 (2015).
14. The process is beautifully described in *Sapiens*, Chapter 5.

15. *Sapiens*, p. 114.
16. For a good account of Descartes' philosophy, see *Stanford Encyclopedia of Philosophy*, 'René Descartes', SEP (2008, update 2014).
17. Raj Patel and Jason W. Moore, *A History of the World in Seven Cheap Things: A Guide to Capitalism, Nature, and the Future of the Planet*, University of California (2019), p. 51.
18. *The Human Planet*, p. 197.
19. This narrative about von Humboldt is inspired by Andrea Wulf, *The Invention of Nature: The Adventures of Alexander von Humboldt, the Lost Hero of Science*, John Murray (2015).
20. Hans Rosling, *Factfulness: Ten Reasons We're Wrong About the World – And Why Things Are Better Than You Think*, Sceptre (2018), pp. 52, 60, 61.
21. Will Steffen et al., 'The trajectory of the Anthropocene: The Great Acceleration', *The Anthropocene Review*, Vol. 2, Issue 1 (2015).
22. Jurriaan M. De Vos et al., 'Estimating the normal background rate of species extinction', *Conservation Biology*, Vol. 29, Issue 2 (2014).
23. Living Planet Report – *2018: Aiming Higher*, WWF (2018).
24. Tim Newbold et al., 'Has land use pushed terrestrial biodiversity beyond the planetary boundary? A global assessment', *Science*, Vol. 353 (2016), pp. 288–91. See also: Chaplin-Kramer et al., 'Global modeling of nature's contributions to people', *Science*, Vol. 366 (2019), pp. 255–8.
25. 'Until and unless we can bring biodiversity back up, we're playing ecological roulette', in: Adam Vaughan, 'Biodiversity is below safe levels across more than half of world's land – study', *Guardian* (14 July 2016).
26. The same point is made by Elizabeth Kolbert, *The Sixth Extinction: An Unnatural History*, Picador (2014), p. 269.
27. Colin N. Waters, 'The Anthropocene is functionally and stratigraphically distinct from the Holocene', *Science*, Vol. 351 (2016).

2. There's Something in the Air

1. My example is inspired by Bill McKibben, *The End of Nature*, Random House (2006), p. 5.
2. I'm calculating based on dry air here. If humidity increases, there is more vapour in the air and the ratios are somewhat different. See: Pieter Tans and Kirk Thoning, 'How we measure background CO_2 levels on Mauna Loa', *NOAA Earth System Research Laboratory* (2008, update 2018).
3. The French scientist Joseph Fourier was one of the first to describe this.
4. The female researcher Eunice Foote was ahead of Tyndall: as early as 1856 she wrote about the warming effects of gases such as CO_2. When Tyndall published his findings, he ignored those of Foote. Tyndall continues to get all the credit, which isn't entirely unfair: Foote's

experiments didn't provide conclusive proof. Further information: Akshat Rathi, 'The female scientist who identified the greenhouse-gas effect never got the credit', *Quartz* (14 May 2018).

5. In fact, without greenhouse gases it would be even colder than minus 18 degrees Celsius, because the white surface of an icy planet reflects more of the sun's rays. If you look exclusively at the presence or absence of greenhouse gases, that gives you a difference of around 33 degrees Celsius.
6. I have written a profile of one of the researchers who works on this: 'Puzzelen met de prehistorie' ('Wrangling with the puzzles of prehistory'), *De Groene Amsterdammer* (23 September 2014).
7. Jeremy D. Shakun et al., 'Global warming preceded by increasing carbon dioxide concentrations during the last deglaciation', *Nature*, Vol. 484 (2012). For a detailed description of the processes that play a role in the transition from ice ages to interglacial periods, see: P. Köhler and H. Fischer, 'Simulating low frequency changes in atmospheric CO_2 during the last 740,000 years', *Climate of the Past*, Vol. 2 (2006), pp. 57–78. And: James E. Hansen and Makiko Sato, 'Paleoclimate Implications for Human-Made Climate Change', *Climate Change* (2012), pp. 21–47.
8. At a CO_2-concentration of under 240 ppm the ice caps would have started to grow thousands of years ago. Source: A. Ganopolski et al., 'Critical insolation–CO_2 relation for diagnosing past and future glacial inception', *Nature*, Vol. 529 (2016), p. 201.
9. A. Berger et al., 'An Exceptionally Long Interglacial Ahead?', *Science*, Vol. 297, Issue 5,585 (2002), p. 1,287. See also: *The Human Planet*, p. 141.
10. I have borrowed the image of crumbling mountains from *Origins*, p. 42. Weathering of stone currently eliminates around 1 billion tonnes of CO_2 per year, see: Jessica Strefler et al., 'Potential and costs of carbon dioxide removal by enhanced weathering of rocks', *IOP Science*, Vol. 13, Issue 3 (2018).
11. The critical threshold for the beginning of a new ice age is probably at a CO_2-concentration of 240 ppm: P. C. Tzedakis et al., 'Determining the natural length of the current interglacial', *Nature Geoscience*, Vol. 5 (2012), pp. 138–41.
12. William F. Ruddiman, 'The Anthropogenic Greenhouse Era Began Thousands of Years Ago', *Climatic Change*, Vol. 61, Issue 3 (2003), pp. 261–93.
13. Ruddiman describes things beautifully in his book *Plows, Plagues, and Petroleum: How Humans Took Control of Climate*, Princeton University Press (2005).
14. Here I use the 'warming potential' of methane over a period of twenty

years. I've chosen this period (instead of the customary hundred years) because methane breaks down in the atmosphere after an average of nine years. Source: IPCC AR5 WG1, Chapter 8, p. 731.

15. My own calculation based on the 'radiative forcing' of different greenhouse gases according to the IPCC between 1750 and 2011: IPCC AR5 WG1, Chapter 8, p. 678, table 8.2.
16. For a critique of Ruddiman's hypothesis, see: B. D. Stocker et al., 'Sensitivity of Holocene atmospheric CO_2 and the modern carbon budget to early human land use: analyses with a process-based model', *Biogeosciences*, Vol. 8 (2011), pp. 69–88. For a recent overview in favour of Ruddiman's hypothesis, see: W. F. Ruddiman et al., 'Late Holocene climate: Natural or anthropogenic?', *AGU100* (2015).
17. This deviation of 0.7 degrees Celsius is known as the 'Little Ice Age': Shaun A. Marcott et al., 'A Reconstruction of Regional and Global Temperature for the Past 11,300 Years', *Science*, Vol. 339, Issue 6,124 (2013), pp. 1,198–1,201.
18. For a good description of this process, see *Origins*, Chapter 9.
19. Eric Monnin et al., 'Atmospheric CO_2 Concentrations over the Last Glacial Termination', *Science*, Vol. 291, Issue 5,501 (2001), pp. 112–14.
20. Lava flows at speeds between 0.1 and 2.7 metres per second, so on average 1.4 metres per second. A skydiver in free fall, belly down, falls at 54 metres per second. A skydiver in free fall therefore moves at fifty times the speed of slow-flowing lava. Source: Wolfram | Alpha.
21. My calculation is based on the 'radiative forcing' of different greenhouse gases according to the IPCC between 1750 and 2011: IPCC AR5 WG 1, Chapter 8, p. 678, table 8.2.
22. Richard E. Zeebe, Andy Ridgwell and James C. Zachos, 'Anthropogenic carbon release rate unprecedented during the past 66 million years', *Nature Geoscience*, Vol. 9 (2016), pp. 325–9.
23. Under the current policy we're hurtling towards 2.3 to 4.1 degrees of warming, to be precise. I'm basing my arguments here on the prognoses of Climate Action Tracker, in which the policy ambitions of countries are worked out as well as possible: 'The CAT Thermometer', *CAT* (December 2019, consulted 20 January 2020).
24. 'How we measure background CO_2 levels on Mauna Loa' (2008, update 2018).
25. Neela Banerjee et al., 'Exxon's Own Research Confirmed Fossil Fuels' Role in Global Warming Decades Ago', *Inside Climate News* (16 September 2015).
26. That became clear for example in 2010, when the IPCC cited incorrect

information from a non-scientific source: Fred Pearce, 'Debate heats up over IPCC melting glaciers claim', *New Scientist* (8 January 2010).

27. Warren Cornwall, 'Even 50-year-old climate models correctly predicted global warming', *Science Magazine* News (4 December 2019).
28. John Cook et al., 'Consensus on consensus: a synthesis of consensus estimates on human-caused global warming', *IOP Science*, Vol. 11, Issue 4 (2016). And: James Lawrence Powell, 'Climate Scientists Virtually Unanimous: Anthropogenic Global Warming Is True', *Bulletin of Science, Technology & Society*, Vol. 35, Issue 5–6 (2016). And: 'Scientific consensus: Earth's climate is warming', NASA.
29. Benjamin D. Santer et al., 'Celebrating the anniversary of three key events in climate change science', *Nature Climate Change*, Vol. 9 (2019), pp. 180–82.
30. This image is nicely outlined by: John Berger, *Pig Earth*, First Vintage International Edition (1992), p. xix.
31. Zadie Smith writes beautifully about this longing for innocence in: 'Elegy for a Country's Seasons', *New York Review of Books* (3 April 2014).
32. Pierre Friedlingstein et al., 'Global Carbon Budget 2019', *Earth System Science Data*, Vol. 11, Issue 4 (2019), pp. 1,783–1,838.
33. *Climate Change 2014: Synthesis Report. Contribution of Working Groups I, II and III to the Fifth Assessment Report of the Intergovernmental Panel on Climate Change*, IPPC (2014), p. 46. This report is henceforth referred to as 'AR5 SYR'.
34. 'Global Carbon Budget 2019'.
35. 'Global Carbon Budget 2019'.
36. In total, about 60 per cent of all our emissions since 1750 has disappeared from the atmosphere due to the effect of 'sinks' such as forests and seas: IPCC AR5 SYR.
37. Zhenzhong Zeng, 'Climate mitigation from vegetation biophysical feedbacks during the past three decades', *Nature Climate Change*, Vol. 7 (2017), pp. 432–6.
38. It should be noted that new forests are rather less interesting than primary forests in terms of biodiversity. For example see: Johnny Wood, 'Earth has more trees than it did 35 years ago – but there's a huge catch', *World Economic Forum* (2018).
39. Zaichun Zhu et al., 'Greening of the Earth and its drivers', *Nature Climate Change*, Vol. 6 (2016), pp. 791–5. And: Yude Pan et al., 'A Large and Persistent Carbon Sink in the World's Forests', *Science*, Vol. 333, Issue 6,045 (2011), pp. 988–93.
40. Jiafu Mao et al., 'Human-induced greening of the northern extratropical land surface', *Nature Climate Change*, Vol. 6 (2016), pp. 959–63.

41. Chi Chen et al., 'China and India lead in greening of the world through land-use management', *Nature Sustainability*, Vol. 2 (2019), pp. 122–9. For an accessible summary see: Daisy Dunne, 'One-third of world's new vegetation in China and India, satellite data shows', *Carbon Brief* (12 February 2019).
42. Edward T. A. Mitchard, 'The tropical forest carbon cycle and climate change', *Nature*, Vol. 559 (2018), pp. 527–34.
43. Wenping Yuan et al., 'Increased atmospheric vapor pressure deficit reduces global vegetation growth', *Science Advances*, Vol. 5, No. 8 (14 August 2019).
44. 'The overall boreal forest is transitioning from a net carbon sink to a source.' Corey J. A. Bradshaw and Ian G. Warkentin, 'Global estimates of boreal forest carbon stocks and flux', *Elsevier*, Vol. 128 (2015), pp. 24–30. See also: Daisy Dunne, 'Climate change's impact on soil moisture could push land past "tipping point"', *Carbon Brief* (2019).
45. Bronson W. Griscom et al., 'Natural climate solutions', *PNAS*, Vol. 114, Issue 44 (2017).
46. IPCC AR5 WG 1, Chapter 6, box 6.1, p. 472.
47. The comparison with radioactive waste is made in this article: Mason Inman, 'Carbon is forever', *Nature Climate Change* (2008), pp. 156–8.
48. The costs do fall. Jeff Tollefson, 'Sucking carbon dioxide from air is cheaper than scientists thought', *Nature News* (7 June 2018).
49. *Global Warming of 1.5 °C*. An IPCC Special Report on the impacts of global warming of 1.5 °C above pre-industrial levels and related global greenhouse gas emission pathways, in the context of strengthening the global response to the threat of climate change, sustainable development and efforts to eradicate poverty, IPCC (2018), Chapter 2, p. 158. This report is henceforth referred to as 'SR1.5'.

3. All Bets Are Off

1. WMO confirms 2019 as second hottest year on record, WMO, 15 January 2020.
2. These rankings are based on time series compiled by the American National Oceanic and Atmospheric Administration (NOAA), National Centers for Environmental information, Climate at a Glance: Global Time Series, published March 2020, retrieved on 20 March, 2020.
3. The phenomenon of the constantly fluctuating frame of reference is known as a 'shifting baseline': 'Shifting baseline', Wikipedia (2019).
4. See 'Health Impacts' and 'Economic Impacts' in Camilo Mora et al., 'Broad threat to humanity from cumulative climate hazards intensified

by greenhouse gas emissions', *Nature Climate Change*, Vol. 8 (2018), pp. 1,062–71.

5. Daisy Dunne, 'Mosquitoborne diseases could reach extra "one billion people" as climate warms', *Carbon Brief* (28 March 2019).
6. An example of mortality as a result of such extreme heat: Meher Ahmad, 'Looking for a Bit of Shade as Intense Heat Wave Hits Karachi', *The New York Times* (29 May 2018).
7. Robert McSweeney, 'Billions to face "deadly threshold" of heat extremes by 2100, finds study', *Carbon Brief* (19 June 2017). See also: Camilo Mora et al., 'Global risk of deadly heat', *Nature Climate Change*, Vol. 7 (2017), pp. 501–6.
8. Kristina Dahl et al., 'Killer Heat in the United States: Climate Choices and the Future of Dangerously Hot Days', Union of Concerned Scientists (July 2019).
9. Mike Kendon et al., 'State of the UK climate 2018', *International Journal of Climatology*, Vol. 39, Issue S1 (2019), p. 3.
10. 'UK extreme events – Heavy rainfall and floods', Metoffice.gov.uk.
11. Met Office, National Climate Information Centre, *State of the UK Climate 2017: Supplementary report on Climate Extremes* (2019), p. 3.
12. Saran Aadhar and Vimal Mishra, 'A substantial rise in the area and population affected by dryness in South Asia under 1.5°C, 2.0°C and 2.5°C warmer worlds', *Environmental Research Letters* (2019).
13. Worldwide, the overall surface that burns down each year is falling, because farmers in Asia and Africa are burning less forest and the tropics have become wetter. But in the more moderate and northern regions, the risk of devastating fires has increased, partly as a result of global warming: Daisy Dunne, 'CO_2 emissions from wildfires have fallen over past 80 years, study finds', *Carbon Brief* (17 April 2018).
14. Emma Marris, 'US wildfires: smoke billows and we're stuck indoors – this is how we live now', *Guardian* (6 August 2018).
15. Lijing Cheng et.al., 'Record-Setting Ocean Warmth Continued in 2019', *Advances in Atmospheric Sciences*, Vol. 37, Issue 2, pp. 137–42 (2020).
16. IPCC, 2019: Summary for Policymakers. In: *IPCC Special Report on the Ocean and Cryosphere in a Changing Climate*, p. 10. In press.
17. The IMBIE team, 'Mass balance of the Antarctic Ice Sheet from 1992 to 2017', *Nature*, Vol. 558 (2018), pp. 219–22.
18. See for example researcher Christina Hulbe's remarks in this article: Eric Holthaus, 'Antarctic melt holds coastal cities hostage. Here's the way out', *Grist* (13 June 2018).
19. Eric Rignot et al., 'Four decades of Antarctic Ice Sheet mass balance from 1979–2017', *PNAS*, Vol. 116, Issue 4 (2019).

20. IPCC, 2019: Summary for Policymakers. In: *IPCC Special Report on the Ocean and Cryosphere in a Changing Climate*. In press.
21. Robert M. DeConto et al., 'Contribution of Antarctica to past and future sealevel rise', *Nature*, Vol. 531 (2016), pp. 591–7.
22. IJssmelt Antarctica in volgende eeuw rampzalig ('Melting ice in Antarctica disastrous in the next century'), KNMI (9 May 2016).
23. Peter U. Clark, 'Consequences of twenty-first-century policy for multi-millennial climate and sea-level change', *Nature Climate Change*, Vol. 6 (2016), pp. 360–69.
24. Even if we continue to emit extreme levels of CO_2 and from this day forward none of it is offset or counteracted by natural carbon sinks or by CO_2 scrubbing technology, the maximum contribution to sea-level rise from Antarctica will be roughly 36 metres by the year 7000. See table 1 in Robert M. DeConto and David Pollard, 'Contribution of Antarctica to past and future sealevel rise', *Nature*, Vol. 531 (2016), pp. 591–7.
25. This would happen in the event of 3 degrees of warming above preindustrial temperatures from before 1750: 'Consequences of twenty-first-century policy for multi-millennial climate and sea-level change', p. 365.
26. So writes Bill McKibben in 'How extreme weather is shrinking the planet', *New Yorker* (16 November 2018). His claim is based on Vietnamese research covered in Alex Chapman and Van Pham Dang Tri, 'Climate change is triggering a migrant crisis in Vietnam', *The Conversation* (9 January 2018).
27. Elizabeth Kolbert, 'The Siege of Miami: As temperatures climb, so, too, will sea levels,' *New Yorker* (13 December 2015).
28. Aaron Rupar, 'Trump dismisses the economic impact of climate change – except at his golf course,' *Vox* (27 November 2018).
29. Liangzhi You et al., 'Impact of growing season temperature on wheat productivity in China,' *Agricultural and Forest Meteorology*, Vol. 149, Issue 6–7 (2009), pp. 1,009–14.
30. Senthold Asseng et al., 'The impact of temperature variability on wheat yields,' *Global Change Biology*, Vol. 17, Issue 2 (2011), pp. 997–1,012. For the anticipated drop in harvests, see IPCC AR5 SYR, p. 15, as well as IPCC SR1.5, Chapter 3, p. 236.
31. *IPCC, 2019: Climate Change and Land*: an IPCC special report on climate change, desertification, land degradation, sustainable land management, food security, and greenhouse gas fluxes in terrestrial ecosystems, Chapter 5, p. 463. In press. From now on referred to as 'IPCC, 2019: Climate Change and Land'.
32. Michiel Korthals, *Goed eten: Filosofie van voeding en landbouw* (*Good Food:*

A philosophy of nutrition and agriculture), Vantilt (2018), p. 263. Korthals' research is based on information from the UK's Department for Environment, Food and Rural Affairs (DEFRA).

33. The State of Food Security and Nutrition in the World 2018: Building climate resilience for food security and nutrition, FAO (2018), p. 28.
34. Mora et al., 'Broad threat to humanity from cumulative climate hazards intensified by greenhouse gas emissions'.
35. The International Agency for Research on Cancer concluded in 2015 that the herbicide glyphosate is 'probably carcinogenic': 'Monographs – 112, Glyphosate', IARC.
36. Matt Richtel and Andrew Jacobs, 'A Mysterious Infection, Spanning the Globe in a Climate of Secrecy', *The New York Times* (6 April 2019).
37. Maria A. Tsiafouli et al., 'Intensive agriculture reduces soil biodiversity across Europe', *Global Change Biology* via *Wiley Online Library*, Vol. 21, Issue 2 (2014), pp. 973–85.
38. The net number of pest organisms goes up on intensively farmed soil, as shown by this study and an interview I had with Wim van der Putten, ecologist and head of the department of Terrestrial Ecology at the Netherlands Institute of Ecology (NIOOKNAW): Maria A. Tsiafouli, 'Intensive agriculture reduces soil biodiversity across Europe,' *Global Change Biology*, Vol. 21, Issue 2 (2014), pp. 973–85.
39. E. C. Oerke, 'Crop losses due to pests,' *Journal of Agricultural Science*, Vol. 144, No. 1 (2006), pp. 31–43.
40. This is my own calculation based on the 'radiative forcing' of various greenhouse gases between 1750 and 2011 according to the IPCC: IPCC AR5 WG 1, Chapter 8, p. 678, table 8.2.
41. IPCC AR5 SYR, p. 47. See also Sonja Vermeulen et al., 'Climate change and food systems,' *Annual Review of Environment and Resources*, Vol. 37 (2012), pp. 195–222.
42. United Nations Convention to Combat Desertification, *The Global Land Outlook: First Edition* (2017).
43. Status of the World's Soil Resources: Main Report, FAO (2015).
44. Larry Elliot, 'Coronavirus could push half a billion people into poverty, Oxfam warns', *Guardian* (9 April 2020).
45. Abdi Latif Dahir, '"Instead of Coronavirus, the Hunger Will Kill Us." A Global Food Crisis Looms.', *The New York Times* (22 April 2020).
46. IPCC SR1.5, Chapter 3, p. 224. See also Scott C. Doney et al., 'The Growing Human Footprint on Coastal and OpenOcean Biogeochemistry,' *Science*, Vol. 328, Issue 5,985 (2010), pp. 1,514–15.
47. Robert J. Diaz et al., 'Spreading Dead Zones and Consequences for Marine Ecosystems,' *Science*, Vol. 321, Issue 5,891 (2008), pp. 926–9.

48. Fred Pearce, 'Can the World Find Solutions to the Nitrogen Pollution Crisis?,' *Yale Environment 360* (6 February 2018).
49. Jody J. Wright et al., 'Microbial ecology of expanding oxygen minimum zones', *Nature Reviews Microbiology*, Vol. 10 (2012), pp. 381–94.
50. In April 2019, Friederike Otto of World Weather Attribution, an initiative by climate scientists who use model runs to calculate the contribution of climate change to specific extreme-weather events, explained to me that a dedicated study about the influence of climate change on Cyclone Idai was not available yet, although she hoped it would be soon. She added: 'I would be very surprised if we would not see an increase in the likelihood and intensity of the extreme precipitation that came with the cyclone because of climate change, as that is what we have seen with all other tropical cyclones. Sea-level rise will also have played an important role, but I can't say anything about the wind. That, and by how much the rainfall increased, does need a dedicated study.' For a study on the emergence of more severe storms in the region, see Jennifer Fitchett, 'Recent emergence of CAT5 tropical cyclones in the South Indian Ocean', *South African Journal of Science*, Vol. 114, No. 11/12 (2018). For an overview of previous 'attribution' studies, see Roz Pidcock et al., 'Mapped: How climate change affects extreme weather around the world,' *Carbon Brief* (11 March 2019).
51. 'Cyclone Idai: More than 1.5 million children urgently need assistance across Mozambique, Malawi and Zimbabwe', UNICEF (27 March 2019).
52. US Aid, *Southern Africa Tropical Cyclones Fact Sheet* (12 August 2019).
53. Ron Kwok, 'Arctic sea ice thickness, volume, and multiyear ice coverage: losses and coupled variability (1958–2018)', *Environmental Research Letters*, Vol. 13, No. 10 (2018).
54. *Global Warming of 1.5 °C: Summary for Policymakers*, IPCC (2018), p. 6. From now on this report will be referred to as SR1.5 SPM.
55. Katey Walter Anthony et al., '21st-century modeled permafrost carbon emissions accelerated by abrupt thaw beneath lakes', *Nature Communications*, Vol. 9, Issue 3,262 (2018).
56. Chris Mooney, 'Arctic Cauldron,' *Washington Post* (22 September 2018).
57. Ingmar Nitze et al., 'Guest post: How Arctic lakes accelerate permafrost carbon losses', *Carbon Brief* (6 September 2018).
58. David Spratt and Ian Dunlop, *What Lies Beneath: The understatement of existential climate risk*, Breakthrough Online (2018), p. 25.
59. IPCC, Climate Change 2014: Impacts, Adaptation, and Vulnerability. Part A: Global and Sectoral Aspects. Contribution of Working Group II to the Fifth Assessment Report of the Intergovernmental Panel on Climate Change, Cambridge University Press (2014), Chapter 19, p. 1,079.

60. See IPCC SR1.5, Chapter 2, p. 103.
61. On the understatement of climate risks in current models, see *What Lies Beneath*.
62. This is based on an interview I did on 13 September 2018 with paleoclimatologist Appy Sluijs.
63. Appy Sluijs et al., 'Environmental precursors to rapid light carbon injection at the Palaeocene/Eocene boundary', *Nature*, Vol. 450 (2007), pp. 1,218–21.
64. Scott L. Wing et al., 'Transient Floral Change and Rapid Global Warming at the Paleocene-Eocene Boundary', *Science*, Vol. 310, Issue 5,750 (2005), pp. 993–6.
65. Chen Chen et al., 'Estimating regional flood discharge during Palaeocene-Eocene global warming', *Scientific Reports*, Vol. 8, No. 13,391 (2018).
66. The same point is made in *Origins*, p. 280, with a reference to Francesca A. McInerney and Scott L. Wing, 'The Paleocene-Eocene Thermal Maximum: A Perturbation of Carbon Cycle, Climate, and Biosphere with Implications for the Future', *Annual Review of Earth and Planetary Sciences*, Vol. 39 (2011), pp. 489–516.
67. Will Steffen et al., 'Trajectories of the Earth System in the Anthropocene', *PNAS*, Vol. 115, Issue 33 (2014, update 2018).

4. It's Not Too Late

1. Mario J. Molina and F. S. Rowland, 'Stratospheric sink for chlorofluoromethanes: chlorine atomcatalysed destruction of ozone', *Nature*, Vol. 249 (1974), pp. 810–12.
2. There was an important extension of the Montreal Protocol in 2016. See: The Kigali Amendment to the Montreal Protocol: HFC Phase-down, UN Environment Fact Sheet.
3. David Wallace-Wells also writes this in *The Uninhabitable Earth: Life After Warming*, Crown Publishing Group (2019), p. 33.
4. This aim was established after a large-scale review under the auspices of the UN Climate Secretariat: United Nations FCCC, *Report on the structured expert dialogue on the 2013–2015 review*, UNFCCC (2015).
5. IPCC SR1.5, Chapter 3, p. 191.
6. Tom K. R. Matthews, 'Communicating the deadly consequences of global warming for human heat stress', *PNAS*, Vol. 114, Issue 15 (2017).
7. IPCC SR1.5, Chapter 3, p. 213. See also: Jacob Schewe et al., 'Multimodel assessment of water scarcity under climate change', *PNAS*, Vol. 111, Issue 9 (2014).
8. IPCC SR1.5, Chapter 3, p. 226.
9. R. Warren et al., 'The projected effect on insects, vertebrates, and plants

of limiting global warming to 1.5°C rather than 2°C', *Science*, Vol. 360, Issue 6390 (2018), pp. 791–5.

10. Roger Hallam (Extinction Rebellion) addresses Amnesty International, YouTube (4 February 2019).
11. 'Global warming is likely to reach 1.5°C between 2030 and 2052 if it continues to increase at the current rate.' IPCC SR1.5 SPM, p. 6.
12. In the calculation of the number of series of *Game of Thrones* that could be made in a billion years, I assume George R. R. Martin's writing pace of on average one book every three years. Between 2011 and 2019 (with the exception of 2018), HBO made a season a year based on Martin's books. The series eventually went ahead of the publication of his books.
13. Dennis Wesselbaum and Amelia Aburn, 'Gone with the wind: International migration', *Global and Planetary Change*, Vol. 178, pp. 96–109 (2019). See also: Karen McVeigh, 'Africa is humanitarian "blind spot": the world's top 10 forgotten crises – report', *Guardian* (28 January 2020).
14. IPCC SR1.5 SPM p. 14. And: Johan Rockström et al., 'A roadmap for rapid decarbonization', *Science*, Vol. 355, Issue 6331 (2017), pp. 1,269–71.
15. IPCC SR1.5 SPM, p. 14.
16. Ibid.
17. Joeri Rogelj, Michiel Schaeffer and Bill Hare, 'Timetables for Zero Emissions and 2050 Emissions Reductions: State of the Science for the ADP Agreement', *Climate Analytics* (February 2015), p. 2.
18. *Global Energy Review* 2020, IEA (2020).
19. 'Cut Global Emissions by 7.6 Percent Every Year for Next Decade to Meet 1.5°C Paris Target – UN Report', United Nations Climate Change (26 November 2019).
20. On the chance of overshooting temperature limits: Gernot Wagner and Martin L. Weitzman, *Climate Shock: The Economic Consequences of a Hotter Planet*, Princeton University Press (2015), Chapter 3.
21. This can be inferred from the discussion in 'Geophysical uncertainties: climate and Earth system feedbacks' in IPCC SR1.5, Chapter 2, p. 103.
22. *Emissions Gap Report 2018*, UNEP (2018), p. xiv.
23. Todd Sanford et al., 'The climate policy narrative for a dangerously warming world', *Nature Climate Change*, Vol. 4 (2014), pp. 164–6. And: Oliver Milman, 'Planet has just 5% chance of reaching Paris climate goal, study says', *Guardian* (31 July 2017). And: P. Christensen et al., 'Uncertainty in forecasts of longrun economic growth', *PNAS*, Vol. 115, Issue 21 (2018).
24. As Bill McKibben has written previously: 'This battle is epic and undecided. If we miss the two-degree target, we will fight to prevent a

rise of three degrees, and then four. It's a long escalator down to Hell.' In 'How extreme weather is shrinking the planet'.

25. 'It is easier to imagine an end to the world than an end to capitalism'. This quote is attributed both to the Slovenian philosopher Slavoj Žižek and the political thinker Fredric Jameson. Read more about it in: Mark Fisher, *Capitalist Realism: Is there no alternative?*, Zero Books (2009).
26. Nathaniel Rich, 'Losing Earth: The Decade We Almost Stopped Climate Change', *The New York Times Magazine* (1 August 2018). Now also available in the book: Nathaniel Rich, *Losing Earth: A Recent History*, MCD (2019).
27. Naomi Klein fiercely criticised Rich's article, instead blaming capitalism: Naomi Klein, 'Capitalism Killed Our Climate Momentum, Not "Human Nature"', *The Intercept* (3 August 2018).
28. The best-known advocate of this idea is Friedrich Hayek. He built on the thinking of Adam Smith, who described the metaphor of the 'invisible hand' in 1776 in *The Wealth of Nations*, W. Strahan and T. Cadell.
29. Thomas Oudman and Theunis Piersma, *De ontsnapping van de natuur: Een nieuwe kijk op kennis* (*Escape from nature: a new look at knowledge*), Athenaeum (2018), p. 98.
30. This insight forms the point of departure for the fascinating book by Timothy Morton, *Being Ecological*, Pelican Books (2018).
31. Thomas Dixon, 'Forget cutthroat competition: to survive, try a little selflessness', *Guardian* (25 July 2016).
32. Evolutionary biologists call the mechanism whereby collaboration is passed on 'multi-level selection'. See: David Sloan Wilson, *This View of Life: Completing the Darwinian Revolution*, Pantheon (2019).
33. Hannah Natanson, 'Forget what you may have been told. New study says strangers step in to help 90 percent of the time', *Washington Post* (6 September 2019).
34. Rebecca Solnit, *Hope in the Dark: Untold Histories, Wild Possibilities*, Haymarket Books (2016), p. xvii.
35. Rebecca Solnit, *A Paradise Built in Hell: The Extraordinary Communities That Arise in Disaster*, Penguin Books (2010).
36. Rebecca Solnit, 'Tomgram: Rebecca Solnit, A Shadow Government of Kindness', TomDispatch (21 December 2010).
37. For much more on our innate capacity for kindness, see: Rutger Bregman, *Humankind: A Hopeful History*, Bloomsbury Publishing (2020).
38. George Monbiot, 'The horror films got it wrong. This virus has turned us into caring neighbours', *Guardian* (31 March 2020).

5. Future Scenario 1: Walls

1. *1988 Shell Confidential Report 'The Greenhouse Effect'*, Shell (1988). I made this document public on *Climate Files* (2018).
2. See my article: 'Shell made a film about climate change in 1991 (then neglected to heed its own warning)', *The Correspondent* (28 February 2017).
3. On this topic I've written: 'Waarom maakte Shell een alarmistische film over klimaatverandering?' ('Why did Shell make an alarmist film about climate change?'), *De Correspondent* (28 February 2017).
4. A good review of the oil industry's early knowledge can be found in Carroll Muffett and Steven Feit, 'Smoke and Fumes: The Legal and Evidentiary Basis for Holding Big Oil Accountable for the Climate Crisis', *CIEL* (November 2017). See also Benjamin Franta, 'Early oil industry knowledge of CO_2 and global warming', *Nature Climate Change*, Vol. 8 (2018), pp. 1,024–5.
5. E. Robinson and R. C. Robbins, 'Sources, abundance, and fate of gaseous atmospheric pollutants', *Smoke and Fumes* (1968).
6. This argument was made in the following interview I conducted: 'De Shelldialogen (1): "Ik maak me zorgen over het klimaat en ik loop niet in een spagaat op mijn werk"' ('The Shell Dialogues (1): "I worry about the climate yet I'm not caught in a bind at work"'), *De Correspondent* (5 May 2016).
7. See my piece: 'De Shelldialogen (6): Shellers verschuilen zich achter de hypocrisie van de consument' ('The Shell Dialogues (6): Shell workers are hiding behind consumer hypocrisy'), *De Correspondent* (7 July 2016).
8. I wrote the following article about the history of Shell: 'Reconstructie: Zo kwam Shell erachter dat klimaatverandering levensgevaarlijk is (en ondermijnde het alle serieuze oplossingen)' ('Reconstruction: Here's how Shell discovered that climate change is a danger to life (and undermined all serious solutions)'), *De Correspondent* (28 February 2017).
9. Clifford Krauss, 'How a "Monster" Texas Oil Field Made the U.S. a Star in the World Market', *The New York Times* (3 February 2019).
10. 'Big Oil's Real Agenda on Climate Change', *InfluenceMap* (March 2019).
11. 'Exclusive: No choice but to invest in oil, Shell CEO says,' Reuters (15 October 2019).
12. Margaret Thatcher, Speech to the Royal Society, margaretthatcher.org (27 September 1988).
13. Naomi Klein made this observation in *This Changes Everything: Capitalism vs. The Climate*, Simon & Schuster (2015).
14. Alex Trembath et al., 'Where the Shale Gas Revolution Came From:

Government's Role in the Development of Hydraulic Fracturing in Shale', Breakthrough Institute (May 2012).

15. The 'fossil fuel rents as a percentage of GDP' are extremely high in these countries: Global Commission on the Geopolitics of Energy Transformation, *A New World: The Geopolitics of the Energy Transformation*, IRENA (2019), p. 31.
16. James D. Ward et al., 'Is Decoupling GDP Growth from Environmental Impact Possible?', *PLOS ONE*, Vol. 11, Issue 10 (2016).
17. The same point is made by Roy Scranton, *Learning to Die in the Anthropocene: Reflections on the End of a Civilization*, City Lights Publishers (2015), p. 53.
18. This figure comes from Shell and is cited in *A New World*, p. 64.
19. World Energy Investment 2018, *IEA* (2018).
20. Ma Jun et al., 'Decarbonizing the Belt and Road: A Green Finance Roadmap', Tsinghua University Center for Finance and Development, Vivid Economics and the Climateworks Foundation (2019).
21. Mind the gap: the $1.6 trillion energy transition risk, Carbon Tracker (8 March 2018). See also: *Production Gap Report 2019*, UNEP et al. (November 2019).
22. Elizabeth Bast et al., *G20 governments propping up fossil fuel exploration, ODI* (2014).
23. Wataru Matsumura and Zakia Adam, 'Fossil fuel consumption subsidies bounced back strongly in 2018', *IEA* (13 June 2019).
24. 'Subsidies for renewables in power generation amounted to $140 billion in 2016.' Source: Toshiyuki Shirai, 'Commentary: Fossil fuel consumption subsidies are down, but not out', *IEA* (20 December 2017).
25. David Roberts, 'Friendly policies keep US oil and coal afloat far more than we thought', *Vox* (26 July 2018).
26. Between 2014 and 2016. Ipek Gençsü and Florian Zerzawy, 'Monitoring Europe's fossil fuel subsidies: Germany', ODI (2017).
27. Laurie van der Burg and Matthias Runkel, Monitoring Europe's fossil fuel subsidies: United Kingdom, ODI (September 2017).
28. Special report: Greening: a more complex income support scheme, not yet environmentally effective, European Court of Auditors (2017).
29. 'Modern feudalism' is how *The New York Times* describes the European system of agricultural subsidies. See: 'The Money Farmers: How Oligarchs and Populists Milk the E.U. for Millions' (3 November 2019).
30. Damian Carrington, 'Angela Merkel "blocks" EU plan on limiting emissions from new cars', *Guardian* (28 June 2013) and Michelle Martin and Edmund Blair, 'Germany's Merkel warns against tougher EU CO_2 limits on cars, vans', Reuters (25 September 2018).

31. Poland's position is reflected in the EU's policy, see points 2 (climate targets) and 2.5 (exceptions stipulated by the likes of Poland) in this major resolution: European Council, 'Presidency conclusions', EUCO 169/14, CO EUR 13, CONCL 5, Brussels (24 October 2014).
32. Abrahm Lustgarten, 'Palm Oil Was Supposed to Help Save the Planet. Instead It Unleashed a Catastrophe', *The New York Times Magazine* (20 November 2018).
33. Duncan Brack, 'Woody Biomass for Power and Heat: Impacts on the Global Climate', Chatham House (23 February 2017).
34. This is why the Committee on Climate Change says 'the UK should "move away" from large-scale biomass power plants'. See: Simon Evans, 'CCC: UK should "move away" from large scale biomass burning', *Carbon Brief* (15 November 2018).
35. 'Why current negative-emissions strategies remain "magical thinking"', *Nature* (21 February 2018), and Daisy Dunne, 'Negative emissions have "limited potential" to help meet climate goals', *Carbon Brief* (31 January 2018).
36. See, for example, IPCC SR1.5, Chapter 2, p. 125.
37. This estimate of 3,000 large factories for the capture of CO_2 was taken from *Shell Scenarios – Sky: Meeting the Goals of the Paris Agreement* (2018), Shell. See also Simon Evans, 'In-depth: Is Shell's new climate scenario as "radical" as it says?', *Carbon Brief* (29 March 2018).
38. In a scenario in which we make extensive use of emission compensation, the IPCC estimates that we'd need 724 million hectares for biofuel plantations by 2050. That's more than twice the total surface area of India (328.7 million hectares). Source: IPCC SR1.5 SPM, p. 19.
39. In 2015, the world's total arable land was some 1,591 million hectares: Kees Klein Goldewijk et al., 'New anthropogenic land use estimates for the Holocene – HYDE 3.2', *Earth Syst. Sci. Data Discuss* (2017), pp. 1–40.
40. Simon L. Lewis et al., 'Restoring natural forests is the best way to remove atmospheric carbon', *Nature* (2 April 2019).
41. Anna B. Harper et al., 'Land-use emissions play a critical role in land-based mitigation for Paris climate targets', *Nature Communications*, Vol. 9 (2018). An accessible summary can be found in Anna Harper, 'Guest post: Why BECCS might not produce "negative" emissions after all', *Carbon Brief* (14 August 2018).
42. Lena R. Boysen et al., 'Earth's Future: The limits to global-warming mitigation by terrestrial carbon removal', *AGU100* (2017), p. 470.
43. P. J. Crutzen et al., 'N_2O release from agrobiofuel production negates global warming reduction by replacing fossil fuels', *Atmospheric Chemistry and Physics*, Vol. 8 (2008).

44. James Hansen et al., 'Young people's burden: requirement of negative CO_2 emissions', *Earth System Dynamics* (2017).
45. There's a fitting term for the corresponding risk: 'moral hazard', a situation in which people take bigger risks in the knowledge that they won't suffer the consequences. See Kevin Anderson et al., 'The trouble with negative emissions', *Science*, Vol. 354, Issue 6309 (2016), pp. 182–3.
46. Jonathan Watts and Ben Doherty, 'US and Russia ally with Saudi Arabia to water down climate pledge', *Guardian* (9 December 2018).
47. The history of opposition against climate policy by Republican presidents is summed up in David Roberts, 'The "Trump effect" threatens the future of the Paris climate agreement', *Vox* (12 December 2018).
48. Rebecca Solnit analysed this mentality in 'The Ideology of Isolation,' *Harper's*, July 2016.
49. 'Statement by President Trump on the Paris Climate Accord', White House (1 June 2017).
50. Jim Dobson, 'Billionaire Bunkers: Exclusive Look Inside the World's Largest Planned Doomsday Escape', *Forbes* (12 June 2015).
51. Martin Lukacs, 'New, privatized African city heralds climate apartheid', *Guardian* (21 January 2014).
52. Jeff Goodell, *The Water Will Come*, Little Brown (2017), Chapter 7: 'Walled Cities'.
53. Ed Pilkington, 'A tale of two Irmas: rich Miami ready for tumult as poor Miami waits and hopes', *Guardian* (9 September 2017).
54. Oliver Milman, 'Climate change is making hurricanes even more destructive, research finds', *Guardian* (14 November 2018).
55. Kevin Loria, 'Piles of garbage and debris still litter Florida's streets months after Hurricane Irma – and recovery is still far away', *Business Insider* (14 December 2017).
56. Nancy Klingener, 'Florida Keys Cope With Suicide Spike After Hurricane Irma', *WLRN* (29 July 2018).
57. Camilo Mora et al. (2018), 'Broad threat to humanity from cumulative climate hazards intensified by greenhouse gas emissions'.
58. Robert McSweeney, 'Scientists discuss the role of climate change in the Syrian civil war', *Carbon Brief* (2 March 2015).
59. See 'Security impacts' in Camilo Mora et al. (2018), 'Broad threat to humanity from cumulative climate hazards intensified by greenhouse gas emissions'. See also Somini Sengupta, 'Warming, Water Crisis, Then Unrest: How Iran Fits an Alarming Pattern', *The New York Times* (18 January 2018).
60. 'Global Trends: Forced Displacement in 2017', UNHCR (2018).

61. Bethany Bell, 'Greece migrant crisis: Islanders strike over crowded camps', BBC News (22 January 2020).
62. Militarisation in times of climate change is addressed in this collection of essays: Nick Buxton and Ben Hayes, *The Secure and the Dispossessed*, Pluto Press (2015).
63. 'Approximately 60% of all deaths and disappearances of migrants and refugees worldwide have occurred in the Mediterranean region, followed by Africa with 21% of the total number of deaths.' Source: Mohamad A. Waked, 'The Unwelcomed' (2020), a graphic story based on data from the International Organization for Migration (IOM).
64. International Organization for Migration, 'Migration and Climate Change', *IOM UN Migration*.
65. This analysis is made by Charles C. Mann, 'State of the Species', *Orion Magazine* (2012).
66. The image of 'red zones' and 'green zones' can be found in Naomi Klein, *No is Not Enough*, Penguin Books (2018).
67. The year 2017 was an exceptionally hot one in the US. See Umair Irfan and Brian Resnick, 'Megadisasters devastated America in 2017. And they're only going to get worse', *Vox* (26 March 2018). This report warns of high future costs for the US: *Fourth National Climate Assessment*, US Global Change Research Program (2018). See also 'Counting the cost 2019: a year of climate breakdown', Christian Aid (December 2019).
68. Jeremy Martinich and Allison Crimmins, 'Climate damages and adaptation potential across diverse sectors of the United States', *Nature Climate Change*, Vol. 9 (2019), pp. 397–404, Thomas Sterner, 'Higher costs of climate change', *Nature*, Vol. 527 (2015), pp. 117–78; 'Broad threat to humanity from cumulative climate hazards intensified by greenhouse gas emissions,' and Damian Carrington, 'Tackle climate or face financial crash, say world's biggest investors,' *Guardian* (10 December 2018).
69. Marshall Burke et al., 'Global non-linear effect of temperature on economic production', *Nature*, Vol. 527 (2015), pp. 235–9.
70. Damian Carrington, 'Perfect temperature for economic success – 13C', *Guardian* (21 October 2015).
71. Researchers have speculated about the reasons behind economic contraction at higher temperatures: Chris Mooney, 'Sweeping study claims that rising temperatures will sharply cut economic productivity', *Washington Post* (21 October 2015). For the impact of heat on productivity, see Tord Kjellstrom et al., 'Working on a warmer planet: The effect of heat stress on productivity and decent work', *International Labour Organization* (1 July 2019).
72. Muthukumara Mani et al., *South Asia's Hotspots: The Impact of Temperature*

and Precipitation Changes on Living Standards, World Bank Group (2018), and Somini Sengupta and Nadja Popovich, 'Global Warming in South Asia: 800 Million at Risk', *The New York Times* (28 June 2018).

73. Solomon Hsiang et al., 'Estimating economic damage from climate change in the United States', *Science*, Vol. 356, Issue 6,345 (2017), pp. 1,362–9.
74. 'Poorer' here means 15 to 30 per cent lower gross domestic product (GDP) per capita: Marshall Burke et al., 'Large potential reduction in economic damages under UN mitigation targets', *Nature*, Vol. 557 (2018), pp. 549–53.
75. Naomi Oreskes and Nicholas Stern, 'Climate Change Will Cost Us Even More Than We Think', *The New York Times* (23 October 2019). See also: Ruth DeFries et al., 'The missing economic risks in assessments of climate change impacts', Grantham Research Institute on Climate Change and the Environment (20 September 2019).
76. The same is suggested by the climate concessions made by these countries. See this study: Simon L. Lewis et al., 'Restoring natural forests is the best way to remove atmospheric carbon', *Nature*, Vol. 568 (2019), pp. 25–8, in particular table 2 in the appendix: Simon L. Lewis et al., 'Supplementary information to: Regenerate natural forests to store carbon', *Nature*, Vol. 568 (2019), p. 8.
77. You can read John Lanchester's novel *The Wall* if you're curious what this future might look like for the UK.
78. Marshall Burke et al., 'Climate and Conflict', *Annual Reviews*, Vol. 7 (2015), pp. 577–617, and T. Carleton et al., 'Conflict in a changing climate', *The European Physical Journal Special Topics*, Vol. 225, Issue 3 (2016), pp. 489–511.
79. John Schwartz, 'Greenland's Melting Ice Nears a "Tipping Point," Scientists Say', *The New York Times* (21 January 2019).

6. Future Scenario 2: Forests

1. Victor Court et al., 'Long-Term Estimates of the Energy-Return-on-Investment (EROI) of Coal, Oil, and Gas Global Productions', *Ecological Economics*, Vol. 138 (2017), pp. 145–59. And: Nafeez Ahmed, 'Inside the new economic science of capitalism's slowburn energy collapse', INSURGE Intelligence (21 August 2017).
2. For more differences between renewable and fossil energy, see this paper: Kingsmill Bond, 'Revolution not evolution: Marginal change and the transformation of the fossil fuel industry', University of Oxford (2017), p. 15.
3. *A New World*, p. 23.

4. Simon Evans, 'Solar, wind and nuclear have "amazingly low" carbon footprints, study finds', *Carbon Brief* (8 December 2017).
5. I'm basing this on an interview that my colleague Leon de Korte and I conducted with Auke Hoekstra: 'Nuon bouwt een windpark zonder een cent subsidie. Welkom in een nieuw energietijdperk' ('Nuon is building a wind farm without a single cent of subsidy. Welcome to a new energy era'), *De Correspondent* (20 March 2018).
6. A good book on this technology is by Tim Flannery, *Sunlight and Seaweed: An Argument for How to Feed, Power and Clean Up the World*, Text Publishing (2017).
7. For the rising costs of nuclear energy, see: *Lazard's Levelized Cost of Energy Analysis*, Lazard (2018), p. 7. See also the drama around the building of Hinkley Point in Britain: 'Hitachi's exit puts Britain's nuclear policy in meltdown', *Economist* (24 January 2019).
8. Francois De Beaupuy, 'World's Largest Nuclear Power Producer Confronts Serial Glitches', Bloomberg (31 October 2019).
9. 'Investment in new renewable power capacity was more than twice that of net, new fossil fuels and nuclear power combined, despite large, ongoing subsidies for fossil fuels generation.' *Renewables 2018 Global Status Report*, REN21 (2018).
10. 'Battery Power's Latest Plunge in Costs Threatens Coal, Gas', BloombergNEF (2019).
11. New Energy Outlook 2018, BNEF (2018).
12. I'm basing this on the interview with Auke Hoekstra mentioned above.
13. So speculated, for instance, Ben van Beurden, CEO of Shell Oil: Shadia Nasralla and Ron Bousso, 'Transition to low-carbon energy may accelerate after crisis: Shell', Reuters (30 April 2020).
14. Paul Hawken (ed.), *Drawdown: The Most Comprehensive Plan Ever Proposed to Reverse Global Warming*, Penguin Books (2017), p. 168.
15. See the warnings from Rana Adib, secretary of the renewable energy agency IRENA, in 2018: Marian Willuhn, 'REN21 report: record PV growth but soaring energy demand leaves renewables playing catchup', *PV Magazine* (4 June 2018). See also: 'Tracking Clean Energy Progress: 2017', IEA (2017).
16. Jeremy Leggett, 'Renewables must be scaled up at least six times faster to meet the Paris Agreement target: IRENA', *Future Today* (17 April 2018).
17. James Temple, 'At this rate, it's going to take nearly 400 years to transform the energy system', *Technology Review* (14 March 2018).
18. T. W. Brown at al., 'Response to "Burden of proof: A comprehensive review of the feasibility of 100% renewable electricity systems"', *Renewable and Sustainable Energy Reviews*, Vol. 92 (2018), pp. 834–47.

19. Arnulf Grubler et al., 'A low energy demand scenario for meeting the 1.5°C target and sustainable development goals without negative emission technologies', *Nature Energy*, Vol. 3 (2018), pp. 515–27.
20. On agroecology, read this report: *From Uniformity to Diversity*, iPES Food (2016). See also this blazing plea for the method: *Report of the Special Rapporteur on the right to food*, United Nations (2017).
21. Iain J. Gould et al., 'Plant diversity and root traits benefit physical properties key to soil function in grasslands', *Ecology Letters*, Vol. 19, Issue 9 (2016).
22. The yield per hectare is higher in policultures (multiple, complementary crops in one place) than in monocultures. Source: *From Uniformity to Diversity*, p. 23 and 31, with references to various studies. See also: *Drawdown*, p. 47. With reference to: Harley I. Manner, *Sustainable traditional agricultural systems of the Pacific Islands*, Permanent Agricultural Resource (2014).
23. Cornelia Rumpel et al., 'Put more carbon in soils to meet Paris climate pledges', *Nature*, Vol. 564 (2018), pp. 32–4.
24. John Kerr and John Landry, *Pulse of the Fashion Industry*, Global Fashion Agenda and Boston Consulting Group (2017), p. 77.
25. *Drawdown*, p. 117.
26. IPCC, *2019: Climate Change and Land*, Chapter 2, p. 189 onwards.
27. *Drawdown*, p. 162.
28. Bill McKibben, 'A World at War', *The New Republic* (15 August 2016).
29. Mariana Mazzucato, *The Entrepreneurial State*, Ingram Publisher (2015), pp. 157–8.
30. Goksin Kavlak et al., 'Evaluating the causes of cost reduction in photovoltaic modules', *Energy Policy*, Vol. 123 (2018), pp. 700–710. Summary: David Roberts, 'What made solar panels so cheap? Thank government policy', *Vox* (28 December 2018).
31. Craig Morris, 'How Germany helped bring down the cost of PV', *Energy Transition* (20 January 2016).
32. Justin Gillis, 'Sun and Wind Alter Global Landscape, Leaving Utilities Behind', *The New York Times* (13 September 2014).
33. Source for prevented emissions in Germany: 'Indicator: GHG emissions avoided through the use of renewables', German Environment Agency (2019). Compared with emissions in Japan in 2018 (source: Global Carbon Budget, 2019). On falling emissions in the German energy supply, see: Arne Jungjohann and Craig Morris, *The German Coal Conundrum: The status of coal power in Germany's energy transition*, Heinrich Böll Foundation (2014). And: 'Klimabilanz 2018: 4,5 Prozent weniger Treibhausgasemissionen Umwelt bundesamt legt erste detaillierte

Schätzung vor' ('Carbon footprint 2018: 4.5 per cent less greenhouse gas emissions, Environment Agency presents first detailed estimation'), German Environment Agency (2019).

34. *The Entrepreneurial State.*
35. Ivetta Gerasimchuk et al., 'Stories from G20 countries: shifting public money out of fossil fuels', ODI (November 2018).
36. *State and Trends of Carbon Pricing 2019*, World Bank Group (2019).
37. Kelly Beaver, 'Two thirds of Britons believe climate change as serious as coronavirus and majority want climate prioritised in economic recovery', Ipsos Mori (22 April 2020).
38. *Renewables 2018 Global Status Report*, REN21 (2018).
39. Christine Shearer et al., *Boom and Bust 2019: Tracking the Global Coal Plant Pipeline*, Global Energy Monitor (2019).
40. For an up-to-date overview of China's climate policy, see: 'China', *Climate Action Tracker* (2019).
41. Max Hall, 'Solar made up 50% of India's new power capacity in 2018', *PV Magazine* (28 February 2019).
42. Michael Safi, 'India's huge solar ambitions could push coal further into shade', *Guardian* (30 June 2018).
43. Christine Shearer, 'Guest post: How plans for new coal are changing around the world', *Carbon Brief* (13 August 2019). See also the data from The Global Coal Exit List.
44. 'Asia's coal developers feeling left out by cold shoulder from banks', Reuters (25 June 2019).
45. Tom Prater, 'Analysis: Global coal power set for record fall in 2019', *Carbon Brief* (25 November 2019).
46. IEA, 'Global Co2 emissions in 2019' (11 February 2020).
47. *Global trends in renewable energy investment*, BloombergNEF (2018), p. 11, and *Global trends in renewable energy investment*, BloombergNEF (2019), p. 23, Figure 12.
48. Mark Roelfsema, 'With local action, major economies can get closer to meeting Paris climate targets', PBL Netherlands Environmental Assessment Agency (30 August 2018).
49. '173 RE100 companies have made a commitment to go "100% renewable". Read about the actions they are taking and why', RE100 (consulted 21 April 2019).
50. Romy van der Burgh and Evert de Vos, 'De duurzame strategie van Philips: De Stille Groene Reus' ('Philips' sustainable strategy: The Silent Green Giant'), *De Groene Amsterdammer* (11 July 2018).
51. *Annual report 2018: All-time high results and strategic progress*, Ørsted (2019).
52. *Annual Report and Form 20-F 2018*, Shell (2018), p. 24.

53. Joshua S. Hill, 'Siemens Again Tops Clean200 List As Clean Stocks Outperform Fossil Fuels', *Clean Technica* (19 February 2018).
54. According to the IPCC, the risk of passing tipping points is 'high' above 2.5 degrees Celsius, see: IPCC SR1.5, Chapter 3, p. 181.
55. Jens Günther et al., *Den Weg zu einem treibhausgasneutralen Deutschland ressourcenschonend gestalten (A resource-efficient pathway towards a greenhouse gas-neutral Germany)*, German Environment Agency (2019).
56. J. Mercure et al., 'Macroeconomic impact of stranded fossil fuel assets', *Nature Climate Change*, Vol. 8 (2018), p. 592.
57. As one would expect, the losses in the EU coal sector are ongoing. Florence Schulz, 'EU coal power plants incurred €6.6 billion losses in 2019, study reveals', *Euractiv* (24 October 2019).
58. Helen Mountford et al., *Unlocking the Inclusive Growth Story of the 21st Century*, The Global Commission on Economy and Climate (2018). See also: *World Employment and Social Outlook 2018: Greening with Jobs*, International Labour Organization (May 2018).
59. Lucien Georgeson and Mark Maslin, 'Estimating the scale of the US green economy within the global context', *Palgrave Communications*, Vol. 5, No. 121 (2019).
60. Katharine Ricke et al., 'Country-level social cost of carbon', *Nature Climate Change*, Vol. 8 (2018), pp. 895–900. And: 'The costs of climate inaction', *Nature* (25 September 2018).
61. *Energy Access Outlook 2017: World Energy Outlook Special Report*, IEA (2017).
62. Lucas Chancel et al., *Carbon and inequality: from Kyoto to Paris*, Paris School of Economics (2015).
63. *Factfulness*, Chapter 3.
64. If you don't believe that, read the chapter 'Innovations' in *Drawdown*.
65. Maggie Astor, 'No Children Because of Climate Change? Some People Are Considering It', *The New York Times* (5 February 2018).
66. 'The global energy transformation may generate a peace dividend': *A New World*, p. 55, with reference to: Andreas Goldthau et al., 'The Geopolitics of Energy Transformation: Governing the Shift – Transformation Dividends, Systemic Risks and New Uncertainties', *SWP*, Issue 42 (2018).
67. *A New World*, p. 49.
68. Professor of Nature Conservation and Plant Ecology Frank Berendse argues for this in his book: *Wilde apen (Wild apes)*, KNNV Uitgeverij (2016).
69. For a plea to make use of this option, see: George Monbiot, 'Averting Climate Breakdown by Restoring Ecosystems: A call to action', *Natural Climate Solutions*.

70. IPCC SR1.5, Chapter 2, pp. 121–2. Some estimations come out even higher: Bronson W. Griscom et al., 'Natural climate solutions', *PNAS*, Vol. 114, Issue 44 (2017).
71. Simon L. Lewis et al., 'Restoring natural forests is the best way to remove atmospheric carbon'.
72. Carly D. Ziter et al., 'Scale-dependent interactions between tree canopy cover and impervious surfaces reduce daytime urban heat during summer', *PNAS*, Vol. 116, Issue 15 (2019).
73. *Drawdown*, p. 90.
74. Richard Burnett et al., 'Global estimates of mortality associated with longterm exposure to outdoor fine particulate matter', *PNAS*, Vol. 115, Issue 38 (2018).
75. Damian Carrington, 'Air pollution cuts two years off global average lifespan, says study', *Guardian* (20 November 2018).
76. Jonathan Watts, 'Clean air in Europe during lockdown leads to 11,000 fewer deaths', *Guardian* (30 April 2020).
77. Bjørn Grinde and Grete Grindal Patil, 'Biophilia: Does Visual Contact with Nature Impact on Health and WellBeing?', *Public Health*, Vol. 6 (2009), pp. 2,332–43.
78. Bin Jiang et al., 'A Dose-Response Curve Describing the Relationship Between Urban Tree Cover Density and Self-Reported Stress Recovery', *SAGE Journals*, Vol. 48, Issue 4 (2014). See also: *Sour mood getting you down? Get back to nature*, Harvard Health Publishing (July 2018).
79. See for example: Todd C. Frankel and Peter Whoriskey, 'Tossed aside in the "white gold" rush', *Washington Post* (19 December 2016).
80. *Green Technology Choices: The Environmental and Resource Implications of Low-Carbon Technologies*, UN Environment Programme (2017). And: *Green Energy Choices: the Benefits, Risks and Trade-Offs of Low-Carbon Technologies for Electricity Production*, UN Environment Programme (2016).
81. Christina M. Patricola et al., 'Anthropogenic influences on major tropical cyclone events', *Nature*, Vol. 563 (2018).
82. See: Naomi Klein, *The Battle for Paradise*, Haymarket Books (2018). Online version: 'Puerto Ricans and ultrarich "puertopians" are locked in a pitched struggle over how to remake the island', *The Intercept* (20 March 2018).

7. Alternatives and Opportunities

1. Katherine Bouton, 'Reports From the Hive, Where the Swarm Concurs', *The New York Times* (27 September 2010).
2. This description is inspired by that of Roy Scranton in *Learning to Die*

in the Anthropocene, p. 55. His account is based on Thomas D. Seeley, *Honeybee Democracy*, Princeton University Press (2010). On YouTube, Seeley explains how this works: Cornell University, 'Honeybee Decision Making', YouTube (5 March 2012).

3. Scranton (2015), *Learning to Die in the Anthropocene*, p. 55.
4. This specific combination could be seen on page 18 of *de Volkskrant*, 12 February 2019. On the environmental impact of various types of holiday: E. Eijgelaar et al., *Travelling large in 2016: The carbon footprint of Dutch holidaymakers in 2016 and the development since 2002*, Centre for Sustainability (2017), p. 21.
5. 'France's richest gain most from Emmanuel Macron's tax reforms', *Financial Times* (23 January 2019). See also Joss Garman, 'Macron's mistake: Taxing the poor to tackle climate change', *Politico* (12 July 2018).
6. Eduardo Porter, 'Does a Carbon Tax Work? Ask British Columbia', *The New York Times* (1 March 2016).
7. Josh Siegel, 'Senate Republicans argue that progressives' Green New Deal would be impossible and unaffordable', *Washington Examiner* (11 December 2018).
8. Kendra Pierre-Louis, 'No One Is Taking Your Hamburgers. But Would It Even Be a Good Idea?', *The New York Times* (8 March 2019).
9. David Roberts, 'Fox News has united the right against the Green New Deal. The left remains divided', *Vox* (22 April 2019).
10. I have written at length about the drawbacks of apocalyptic climate stories: 'Waarom we zo vaak zwijgen over het klimaat (en hoe we dat kunnen doorbreken)' ('Why we often remain silent on the climate (and how to break this silence)'), *De Correspondent* (20 October 2015).
11. I have borrowed this idea from George Monbiot. 'The only thing that can replace a story is a story,' he wrote in his book *Out of the Wreckage*, Verso (2017).
12. Naomi Klein, 'Puerto Ricans and ultrarich "puertopians" are locked in a pitched struggle over how to remake the island', *The Intercept* (20 March 2018)
13. I'm indebted to Eva Gladek of *Metabolic* for this insight (I interviewed her on 2 October 2018).
14. *The Automobile Industry Pocket Guide*, European Automobile Manufacturers' Association (ACEA) (2017 and 2018).
15. *Unlocking the Inclusive Growth Story of the 21st Century*, The New Climate Economy (2018). Summary: David Roberts, 'We could shift to sustainability and save $26 trillion. Why aren't we doing it?', *Vox* (6 September 2018).

16. 'UK "needs billions a year" to meet 2050 climate targets', *Guardian* (28 September 2019).
17. Office for National Statistics, 'Overseas travel and tourism: August 2019 provisional results' (29 November 2019).
18. Brian O'Callaghan and Cameron Hepburn, 'Leading economists: Green coronavirus recovery also better for economy', *Carbon Brief* (5 May 2020).
19. Fatih Birol, 'Now is the time to plan the economic recovery the world needs', IEA (27 April 2020).
20. *Global Trends in Renewable Energy Investment 2019,* Frankfurt School-UNEP Centre / BNEF (2019), p. 20.
21. 'U.S. coal consumption in 2018 expected to be the lowest in 39 years', US Energy Information Administration (28 December 2018).
22. '2020 Vision'.
23. Investors will be just fine without fossil fuel shares according to Jeremy Grantham, 'The mythical peril of divesting from fossil fuels', The London School of Economics and Political Science (2018).
24. Toby A. A. Heaps, 'Divestment would have made NY pension fund $22B richer', *Corporate Knights* (4 October 2018).
25. 'Almost $1bn wiped off the value of UK pensions', *Financial Times* (10 October 2015).
26. 'In the past 10 years [...] companies in the S&P 500 energy sector had gained just 2 percent in total. In the same period, the broader S&P 500 nearly tripled.' Source: Andrew Ross Sorkin, 'BlackRock C.E.O. Larry Fink: Climate Crisis Will Reshape Finance', *The New York Times* (14 January 2020).
27. 'Response to "Burden of proof: A comprehensive review of the feasibility of 100% renewable-electricity systems"'.
28. '100% Renewable Electricity Worldwide is Feasible and More Cost-Effective than the Existing System', Energy Watch Group (2017). See also *A New World,* p. 63 with reference to 'Energy Transition Outlook 2018', DNVGL (2018).
29. See for example, Andries F. Hof et al., 'Global and regional abatement costs of Nationally Determined Contributions (NDCs) and of enhanced action to levels well below 2°C and 1.5°C', *Environmental Science & Policy*, Vol. 71 (2017), pp. 30–40; Damian Carrington, 'Hitting toughest climate target will save world $30tn in damages, analysis shows', *Guardian* (23 May 2018); and Tom Kompas, 'Earth's Future', *AGU100* (2018).
30. LSE and Grantham Institute, 'The Economics of Climate Change: The Stern Review', LSE (2006). Stern later revised this upwards: Robin McKie, 'Nicholas Stern: cost of global warming "is worse than I feared"', *Guardian* (6 November 2016).

31. Similarly, the *Economist* reports: 'Actuaries calculate that governments investing $1 in climate resilience today will save $5 in losses tomorrow.' See *Economist*, 'One way or another the deluge is coming' (17 August 2019).
32. A lot has been written about this so-called 'resource curse' – see Wikipedia, or my piece (in Dutch): 'Dit is de vloek van olie: als je er eenmaal afhankelijk van bent, dan blijf je dat' ('This is the curse of oil: Once you're hooked, you can't quit'), *De Correspondent* (28 June 2017).
33. I've written an article on this: 'In Ecuador wandelen de jaguars tussen de pijpleidingen. Waarom?' ('In Ecuador the jaguars walk among the pipelines. Why?'), *De Correspondent* (29 June 2017).
34. This example is also cited by David Wallace-Wells in his book *The Uninhabitable Earth*. Source: Amina Khan, 'How much Arctic sea ice are you melting? Scientists have an answer', *Los Angeles Times* (3 November 2016).
35. Jeremy Grantham, 'The Race of our Lives Revisited', *White Paper* (2018), p. 3.
36. David Roberts, 'We are almost certainly underestimating the economic risks of climate change', *Vox* (9 June 2018).
37. 'The Great Gas Lock-in', Corporate Europe Observatory (31 October 2017).
38. 'Billions to be wasted on "unnecessary" gas projects, study says', *Euractiv* (20 January 2020).
39. Jesse Barron, 'How Big Business Is Hedging Against the Apocalypse', *The New York Times Magazine* (11 April 2019).
40. Brian Merchant, 'How Google, Microsoft, and Big Tech Are Automating the Climate Crisis', *Gizmodo* (21 February 2019)
41. Reserve calculations based on *World Energy Outlook 2019*, IEA (2019), p. 754, as well as *Statistical Review of World Energy 2019*, BP (2019), pp. 15 and 31, and Beaudoin et al., *Frozen Heat: A Global Outlook on Methane Gas Hydrates*, UNEP (2014), p. 31.
42. See also 'The U.S. Department of Energy, As Part of an International Team, Has Successfully Drilled a Gas Hydrate Test Well on Alaska North Slope', *Office of Fossil Energy* (23 January 2019).
43. Bruno Latour, *Down to Earth: Politics in the New Climatic Regime*, Polity Press (2018), p. 66.

8. The Battle of the Century

1. Calculation: total emissions from the Netherlands in CO_2 equivalents in 2018 (0.19 gigatonnes) compared with global CO_2 equivalent emissions in 2018 (55.3 gigatonnes). Sources: 'Emissies broeikasgassen, 1990–2018'

('Greenhouse gas emissions, 1990–2018'), CLO (2019). And: 'Emissions Gap Report 2019', UNEP (2019).

2. Ruling: The Hague Court of Appeal, 'Uitspraken' ('Rulings'), Gerechtshof (2018).
3. Cox made these statements in a short film produced by the Dutch documentary programme *Tegenlicht*. Tweet by VPRO *Tegenlicht* (2017).
4. See my article on the subject: 'Het proces: de mensheid versus de Nederlandse staat' ('The trial: humankind against the Dutch state'), *De Correspondent* (10 December 2013).
5. Ruling ECLI:NL:RBDHA:2015:7196, The Hague Court (2015).
6. 'Thanks to this landmark court ruling, climate action is now inseparable from human rights', *The Correspondent* (20 December 2019).
7. The peak can easily be seen in the first graph in this article: 'Climate-change lawsuits', *Economist* (2 November 2017).
8. The status of climate litigation: A global review, UNEP (2017), p. 10. See also: Global trends in climate change legislation and litigation: 2019 snapshot, Grantham Research Institute on Climate Change and the Environment (2019).
9. For an up-to-date overview: 'Juliana v. U.S. Youth Climate Lawsuit', Our Children's Trust.
10. Read more on this case at: 'Climat: stop à l'inaction, demandons justice!' ('Climate: stop the inaction, demand justice'), *L'Affaire du siècle*. And: 'Notre Affaire à Tous and Others v. France', *Climate Case Chart* (2018).
11. Shell acknowledged that responsibility when it announced its own ambition to reduce the 'net carbon footprint' of its products: 'Management Day 2017: Shell updates company strategy and financial outlook, and outlines net carbon footprint ambition', Shell (28 November 2017).
12. See my article: 'Shell moet voor de rechter komen voor zijn aandeel in klimaatverandering. Dit is de advocaat die daarachter zit' ('Shell must be brought before the courts for its share in climate change. This is the lawyer behind the case'), *De Correspondent* (5 April 2019). For all the stories I've written on climate law, see the collection 'Klimaatrechtszaken' ('Climate lawsuits'), *De Correspondent* (since 2013).
13. On Exxon's climate study, see this fantastic journalistic project: Neela Banerjee et al., 'Exxon's Own Research Confirmed Fossil Fuels' Role in Global Warming Decades Ago', *Inside Climate News* (16 September 2015).
14. Neela Banerjee et al., 'Exxon Believed Deep Dive Into Climate Research Would Protect Its Business', *Inside Climate News* (17 September 2015).
15. Geoffrey Supran and Naomi Oreskes, 'Assessing ExxonMobil's climate change communications (1977–2014)', *IOP Science*, Vol. 12, Issue 8 (2017).

16. See: Union of Concerned Scientist, 'The Climate Deception Dossiers: Internal Fossil Fuel Industry Memos Reveal Decades of Corporate Disinformation' (29 June 2015).
17. A good example of such evidence against the fossil industry can be found in the complaint by Sher Edling LLP on behalf of the Pacific Coast Federation of Fishermen's Associations against thirty fossil energy companies.
18. Umair Irfan, 'Pay attention to the growing wave of climate change lawsuits', *Vox* (10 April 2019).
19. Paul Griffin, *The Carbon Majors Database: CDP Carbon Majors Report 2017*, CDP (2017).
20. IRENA makes this claim based on BP's *Statistical Review of World Energy 2018* in *A New World*, p. 82.
21. IPCC SR1.5 SPM, p. 14. And: IPCC SR1.5, Chapter 2, p. 108, table 2.2.
22. In September 2015, Arabella Advisors counted divestment commitments from forty-three countries: *Measuring the Growth of the Global Fossil Fuel Divestment and Clean Energy Investment Movement*, Arabella Advisors (2015).
23. I described the status of the divestment campaign from the climate summit in Paris: 'Dankzij deze activisten werd het klimaatverdrag van Parijs een succes' ('Thanks to these activists the Paris climate agreement was a success'), *De Correspondent* (15 December 2015). At the time, 350.org claimed that the number of divestment commitments had risen above 500: 'Momentum: Campaigners Announce at Paris Climate Summit that Divestment Commitments Have Passed 500+ Institutions Representing $3.4 trillion in Assets', 350.org (2 December 2015).
24. *The Global Fossil Fuel Divestment and Clean Energy Investment Movement – 2018 Report*, Arabella Advisors (2018).
25. 'World's top three asset managers oversee $300bn fossil fuel investments', *Guardian* (12 October 2019).
26. 'Shell CEO urges switch to clean energy as plans hefty renewable spending', Reuters (9 March 2017).
27. Nick Cunningham, 'HSBC Advises Clients To Get Out Of Fossil Fuels', Oilprice.com (29 April 2015).
28. 'Lagarde Vows to Put Climate Change on the E.C.B.'s Agenda', *The New York Times* (4 September 2019).
29. 'EU Bank Takes "Quantum Leap" to End Fossil-Fuel Financing', Bloomberg (14 November 2019).
30. 'Climate change: why Sweden's central bank dumped Australian bonds', *The Conversation* (15 November 2019).
31. 'Goldman Sachs Leads Among US Banks with Commitment Not to Fund Arctic Drilling', Sierra Club (15 December 2019).

32. 'Failure of Aramco IPO Gives a Black Eye to Bank CEOs and a Warning to Investors', Center for International Environmental Law (19 November 2019).
33. My description of Blockadia is based on the example of Naomi Klein in *This Changes Everything*, Chapter 9.
34. Kim Bryan, 'People Power versus Carbon Bombs: The Eight AntiCoal Campaigns that Make Staying at 1.5C Possible', 350.org (4 December 2018).
35. See, for example, the remarks from Esperanza Martinez in: Naomi Klein, *This Changes Everything*, Simon & Schuster, p. 304.
36. Julia Carrie Wong and Sam Levin, 'Standing Rock protesters hold out against extraordinary police violence', *Guardian* (29 November 2016).
37. Julia Carrie Wong, 'Dakota Access pipeline: US denies key permit, a win for Standing Rock protesters', *Guardian* (5 December 2016).
38. The involvement of Ocasio-Cortez at Standing Rock and the cohesion between different climate movements is nicely described by: Rebecca Solnit, 'Standing Rock inspired Ocasio-Cortez to run. That's the power of protest', *Guardian* (14 January 2019).
39. I take this insight from Rebecca Solnit. She writes: 'How can you add up the foreclosures and evictions that don't happen, the forests that aren't leveled, the species that don't go extinct, the discriminations that don't occur?' In: 'Iceberg Economies and Shadow Selves', *The Nation* (22 December 2010).
40. United Nations Environment Programme et al., *Protected Planet Report 2018: Tracking progress towards global targets for protected areas*, UNEP World Conservation Monitoring Centre (2018), p. 7.
41. *Global Environment Outlook 6*, UN Environment (2019).
42. Kendall R. Jones et al., 'One-third of global protected land is under intense human pressure', *Science*, Vol. 360, Issue 6,390 (2018), pp. 788–91.
43. 'Green New Deal: Where 2020 Democrats stand', *Washington Post* (consulted 9 January 2020).
44. Frédéric Simon, 'Green Deal will be "our motor for the recovery", von der Leyen says', *Euractiv* (29 April 2020).
45. Michael Grunwald, 'Biden wants a new stimulus "a hell of a lot bigger" than $2 trillion', *Politico* (25 April 2020).

9. The Power of Small Changes

1. Special Eurobarometer 490 on Climate Change, EU Barometer (April 2019).
2. 'Greta Thunberg addressed the COP24 plenary session December 12th', YouTube (2019).

3. 'Klimaatprotest neemt toe: ongeziene opkomst van 35.000 spijbelaars in Brussel' ('Climate protests are escalating: unprecedented number of 35,000 students playing truant in Brussels'), *De Morgen* (24 January 2019).
4. Greta Thunberg on Twitter (2019).
5. Jessica Glenze et al., 'Climate strikes held around the world – as it happened', *Guardian* (15 March 2019).
6. Sandra Laville and Jonathan Watts, 'Across the globe, millions join biggest climate protest ever', *Guardian* (21 September 2019).
7. Rebecca Solnit, 'Thank you, climate strikers. Your action matters and your power will be felt', *Guardian* (15 March 2019).
8. This phenomenon is known as the shifting 'Overton window'.
9. Jillian Ambrose, 'Fracking halted in England in major government U-turn', *Guardian* (2 November 2019).
10. I'm quoting from the website of the *Convention Citoyenne pour Le Climat* (The Citizen's Assembly for the Climate).
11. Emma Gatten, 'Citizens Assembly lets public consider how UK should meet climate change goals', *Telegraph* (25 January 2020).
12. Brad Plumer, 'It's not just solar panels. Electric cars can be contagious, too', *Vox* (29 August 2016).
13. 'Food and agriculture data', *FAO* (consulted on 20 April 2019).
14. P. J. Gerber et al., *Tackling Climate Change through Livestock: A Global Assessment of Emissions and Mitigation Opportunities*, *FAO* (2013). See also: Francesco N. Tubiello et al., 'The Contribution of Agriculture, Forestry and other Land Use activities to Global Warming, 1990–2012', *Global Change Biology* (2015), and Walter Willett et al., 'Food in the Anthropocene – the EAT-Lancet Commission on healthy diets from sustainable food systems', *The Lancet*, Vol. 393, Issue 10,170 (2019), pp. 447–92.
15. 'Food in the Anthropocene', p. 9. Recent research that suggests otherwise has been discredited. See: 'New "guidelines" say continue red meat consumption habits, but recommendations contradict evidence', Harvard School of Public Health (30 September 2019).
16. 'Food in the Anthropocene', p. 14.
17. This is the diet described in: 'Food in the Anthropocene'.
18. Marco Springmann, 'Analysis and Valuation of the Health and Climate Change Cobenefits of Dietary Change', *PNAS*, Vol. 113, Issue 15 (2016), p. 4,147.
19. Janet Ranganathan et al., *Shifting Diets for a Sustainable Food Future*, World Resources Institute (2016).
20. *Meat Substitutes Market worth 6.43 Billion USD by 2023*, Markets and

Markets. See also Emiko Terazono, 'Investors seek sustenance in alternative proteins', *Financial Times* (4 September 2019).
21. 'Analysis and Valuation of the Health and Climate Change Cobenefits of Dietary Change', pp. 4,148–9.
22. Yinon M. Bar-on et al., 'The biomass distribution on Earth', *PNAS*, Vol. 114 (2018), pp. 6,506–11.
23. '9 out of 10 people worldwide breathe polluted air, but more countries are taking action', *WHO* (2 May 2018).
24. I did these calculations for the article: 'Toen ik deze cijfers onder ogen zag, besloot ik veel minder te vliegen (en jij misschien ook)' ('When I saw these figures, I decided to stop flying (and maybe you will too)'), *De Correspondent* (13 June 2018).
25. The emissions of a Boeing 747 were calculated on Twitter by Kees van der Leun, strategy consultant with Navigant. Tweet from @Sustainable2050 (21 September 2018).
26. I've also derived this analogy from Kees van der Leun: Tweet from @Sustainable2050 (24 March 2019).
27. Niko Kommenda, '1% of English residents take one-fifth of overseas flights, survey shows', *Guardian* (25 September 2019).
28. Hiroko Tabuchi and Nadja Popovich, 'How Guilty Should You Feel About Flying?', *The New York Times* (17 October 2019).
29. This point is made by David Wallace-Wells, *The Uninhabitable Earth: Life After Warming*, Tim Duggan Books (2019). His argument is based on *Extreme Carbon Inequality*, Oxfam (2015).
30. *Drawdown*, p. 42.
31. Detlef P. van Vuuren et al., 'Alternative pathways to the 1.5°C target reduce the need for negative emission technologies', *Nature Climate Change*, Vol. 8 (2018), pp. 391–7.
32. Laura Cozzi and Apostolos Petropoulos, 'Commentary: Growing preference for SUVs challenges emissions reductions in passenger car market', *IEA* (15 October 2019).
33. Rebecca Solnit, '"The impossible has already happened": what coronavirus can teach us about hope', *Guardian* (7 April 2020).
34. Jonathan Watts, 'How Greta Thunberg's school strike went global', *Guardian* (14 March 2019).
35. I derive this insight from the following article: Freek Vielen, 'Alles hangt af van wat wij beslissen, denken, dromen, durven, willen, zeggen, doen' ('Everything depends on what we decide, think, dream, dare, want, say, do'), *De Correspondent* (6 August 2016).
36. Antjie Krog, 'Deze Afrikaanse filosofie inspireert tot een nieuwe soort

verbondenheid' ('This African philosophy inspires a new kind of unity'), *De Correspondent* (15 July 2017).

Appendix

1. Richard Davy et al., 'Diurnal asymmetry to the observed global warming', *International Journal of Climatology*, Vol. 37, Issue 1 (2016).
2. IPCC AR5 WG 1, Working group 1, Chapter 8, pp. 688–90. And: Philip Duffy et al., 'Solar variability does not explain late-20th-century warming', *Physics Today* (2009).
3. Global Trends in Renewable Energy Investment 2019, Frankfurt School-UNEP Centre/BNEF (2019), p. 14.
4. 'International poll: most expect to feel impact of climate change, many think it will make us extinct', YouGov (15 September 2019).
5. Jacob Poushter and Christine Huang, 'Climate Change Still Seen as the Top Global Threat, but Cyberattacks a Rising Concern', PEW Research Centre (10 February 2019).
6. Special Eurobarometer 490 on Climate Change, EU Barometer (April 2019).
7. David Roberts, 'New global survey reveals that everyone loves green energy – especially the Chinese', *Vox* (21 November 2017).
8. 'The Cost of Doing Nothing', International Federation of Red Cross and Red Crescent Societies (2019).
9. TEDxBoulder, 'The success of nonviolent civil resistance: Erica Chenoweth at TEDxBoulder', YouTube (2013). See also: Erica Chenoweth and Maria J. Stephan, *Why Civil Resistance Works: The Strategic Logic of Nonviolent Conflict*, Columbia University Press (2012).

Afterword and Further Reading

1. Tom Lovejoy et al., 'Amazon forest to savannah tipping point could be far closer than thought', *Mongabay* (5 March 2018).
2. This is beautifully described in: Jeff Goodell, *The Water Will Come*, p. 138.
3. The calculation is based on IPCC AR5, Chapter 12, table 12.3. The table gives the expected future temperature rises in different years at different concentrations of CO_2 in the atmosphere, as compared with the average temperatures in the period 1986–2005. The IPCC even states that a rise of 0.61 °C took place *before* 1986. I've added this figure to the IPCC prediction to give a figure representing total warming since the Industrial Revolution.
4. Hanna Bervoets, 'Het is 2075. De klimaatbeweging heeft gewonnen' ('It's 2075. The climate movement has won'), *De Correspondent* (4 October 2015).

ACKNOWLEDGEMENTS

A journalist is nothing without sources. I would like to sincerely thank everyone who has shared his or her knowledge and experience with me in recent years. Special thanks to the people who worked with me on the series of interviews just before I went on leave to write this book: Kees van der Leun, Willem Schinkel, Corinne Le Quéré, Eva Gladek, Appy Sluijs, Marjan Minnesma, Michael Mann, Bas Eickhout, May Boeve, Harriet Bergman and Anneke Wensing. Thanks to my colleagues at *De Correspondent* for countless conversations on the platform: you constantly sharpened my thoughts. Thanks also to my former colleagues at political and cultural weekly *De Groene Amsterdammer* and platform for investigative journalism *Investico*: some of the insights in this book date back to our years of collaboration.

Thanks to the clever suggestions of editor Harminke Medendorp, this book improved with every round of editing. Harminke, many thanks for your support and commitment.

Andreas Jonkers, thanks so much for your boundless efforts, your final edit and sticking to the deadline – your decisiveness helped me enormously. Lisa Roggeveen, thanks for tidying up the endnotes and Riffy Bol for double-checking the calculations.

Many thanks to my fellow readers and thinkers, who constantly helped me improve this book. From *De Correspondent*: Lynn Berger, Rutger Bregman, Jesse Frederik, Joris Luyendijk, Nina Polak, Tamar Stelling, Tomas Vanheste, Maite Vermeulen, Johannes Visser and Rob Wijnberg.

A number of external experts read and commented on early passages of this book. For that I would like to thank Lisette van Beek, Boele Bonthuis, Bart Crezee, Hans Custers, Jos Hagelaars, Maarten Hajer, Bob Klein Lankhorst, Thomas Oudman, Peter Pelzer, Appy Sluijs, Pier Vellinga and Bart Verheggen. Many thanks, also, to Steven Schoon and Dirk Vis for their feedback on the manuscript, and Philip Huff for his thorough final edit. Without you it would all have come to nothing.

I'd like to thank Leon Postma for the fantastic Dutch cover design, Harald Dunnink for his involvement with it, and Steve Panton for the English version.

Jorris Verboon worked with Leon Postma and Leon de Korte to create the pretty infographics – they make it look easy.

Annelieke Tillema, many thanks for your careful corrections of the Dutch manuscript and proofs.

This English translation could not have come about without the mediation of Rebecca Carter from Janklow & Nesbit. She put me in contact with Helen Conford, my publisher at Profile, who has improved this English version at every turn. Many thanks to her and everyone at Profile for all your efforts. Special thanks to translators Laura Vroomen and Anna Asbury for the pleasant collaboration and your dedication to this book.

Dear friends and family, thank you for your patience and support.

And the most thanks, finally, to my beloved Milou Klein Lankhorst, who has played a crucial double role in the making of this book. You're not only the best publisher I could wish for, but also the loveliest person. You were there the whole time, you read everything and helped me at every step. Thank you.